Editor's Choice, *En*

T0032369

"Mr. Vaillant is absolutely spellbinding when conjuring up the world of the golden spruce." —William Grimes, *New York Times*

"A haunting tale of a good man driven mad by environmental devastation. . . . [Grant Hadwin's] appalling tree surgery is as vividly wrought as one of Patrick O'Brian's shipboard amputations."
  —Frank Clifford, *Los Angeles Times*

"This tragic tale goes right to the heart of the conflicts among loggers, native rights activists, and environmentalists, and induces us to more deeply consider the consequences of our habits of destruction."
  —Donna Seaman, *Booklist* (starred)

"In his first book . . . John Vaillant interlaces a well-reported murder mystery with elegantly spun cultural and natural history, conjuring the spooky mood of the Northwest forests with the clarity of David Guterson or Jonathan Raban." —Bruce Barcott, *Outside*

"Vaillant's tale . . . is of unfailing interest." —*Kirkus Reviews*

"Make some more space on the shelf of Essential Northwest Books. John Vaillant has crafted a debut book that is a stunning look at this region's history and environment."
  —John Marshall, *Seattle Post-Intelligencer*

"This book will appeal to anyone interested in the Pacific Northwest, environmentalism, or a gripping real-life mystery." —*Science*

"A page-turner as dramatic as a novel. . . . The story is as majestic as the golden spruce, and we are fortunate to have a writer of Vaillant's exceptional skill to tell the tale."
  —Tom Hawthorn, *Vancouver Sun*

"Readers will be reminded of Jon Krakauer's writing when they pick up John Vaillant's *The Golden Spruce: A True Story of Myth, Madness, and Greed.*" —Susan Swagler, *Birmingham News*

"Writing in a vigorous, evocative style, Vaillant portrays the Pacific Northwest as a region of conflict and violence. . . . Vaillant paints a haunting portrait of man's vexed relationship with nature."
—*Publishers Weekly*

"*The Golden Spruce* is a true crime story replete with a sympathetic victim, a cunning and ambiguous murderer and an unsolved mystery." —Meg Jones, *Milwaukee Journal Sentinel*

"The writing is so vivid that it will make you want to visit the Queen Charlotte Islands. The story is so heartbreaking that it will make you question anew where our civilization is going . . . a gracefully written, ambitiously researched and enthralling story of ecological majesty and human greed . . . a page-turner of a debut."
—William Dietrich, *Seattle Times*

"Add *The Golden Spruce* to your 'short list' of books that best capture an essential element of the Pacific Northwest."
—*Annie Bloom's Books*

"*The Golden Spruce* serves as a highly digestible exploration of the timber industry. . . . Through colorful interviews with present-day loggers, Vaillant reveals the perils of the dwindling, but still-vital, trade." —Katherine Wroth, *Grist Magazine*

"[John Vaillant] tells the story of a man who felled a three-hundred-year-old Sitka spruce in protest of old-growth clearcutting, and tells it on paper that could well have come from a Sitka spruce. Painfully aware of this inconsistency, he writes as if he's determined to produce a work worthy of sacrificing the trees required to print it. He succeeds."   —Louise Redd, *Orion*

"[John Vaillant] gives the reader arresting descriptions of an exotic landscape, along with illuminating discourses on plant genetics, the timber business, and the clash between native culture and corporate capitalism. His book is one to ponder and savor."
                                    —Laurence A. Marschall, *Natural History*

"The case of the golden spruce is a fascinating eco-mystery that cuts right to the heart of modern settlement of the Pacific Northwest."
                                    —Jeff Baker, *The Oregonian*

"[Like] Krakauer and Sebastian Junger, Vaillant deftly peels away the surface layers to explore the psychology below. . . . An intense mystery and a sweeping history, *The Golden Spruce* makes for a terrific read."   —Robert Wiersema, *National Post*

"Vaillant's book is a haunting tale of a good man driven mad by environmental devastation."   —Frank Clifford, *Los Angeles Times*

"Vaillant gives us a vivid tale of mankind's exploitation of the land, a three-way marriage of masculine excess, technological development and unquenchable demand. The telling has a sweaty gusto."
                                    —Thomas H. Rawls, *Minneapolis Star Tribune*

W. W. NORTON & COMPANY *New York London*

THE
GOLDEN
SPRUCE

A True Story of Myth, Madness, and Greed

JOHN VAILLANT

for Nora

Peter Trower, "The Ridge Trees," from *Haunted Hills and Hanging Valleys: Selected Poems 1969–2004*, © Harbour Publishing. Used with permission. Lines 1–3 of "Canto I," from *Inferno: A New Verse Translation* by Dante Alighieri, translated by Michael Palma. Copyright © 2002 by Michael Palma. Used by permission of W. W. Norton & Company, Inc. "Ceremony," from *Ceremony* by Leslie Marmon Silko, copyright © 1977 by Leslie Silko. Used by permission of Viking Penguin, a division of Penguin Group (USA) Inc. Lines from *Faust, Part I* by Johann Wolgang von Goethe, translated by Martin Greenberg, Yale University Press. Copyright © 1992 by Martin Greenberg. Used by permission of Yale University Press. Excerpt from "Un-chopping a Tree" by W. S. Merwin, reprinted with the permission of the Wylie Agency, Inc.

For information about permission to reproduce selections from this book, write to Permissions, W. W. Norton & Company, Inc., 500 Fifth Avenue, New York, NY 10110

Manufacturing by The Maple-Vail Book Manufacturing Group
Book design by Barbara M. Bachman
Cartography by Jacques Chazaud
Production manager: Anna Oler

Library of Congress Cataloging-in-Publication Data

Vaillant, John.
The golden spruce : a true story of myth, madness, and greed / John Vaillant.— 1st ed.
p. cm.
Includes bibliographical references (p.   ).
ISBN 0-393-05887-5 (hardcover)
1. Sitka spruce—British Columbia—Queen Charlotte Islands. 2. Hadwin, Grant. 3. Logging—British Columbia—Queen Charlotte Islands. 4. Haida Indians—British Columbia—Queen Charlotte Islands. I. Title.
SD397.S77V35 2005
333'.7513'0971112—dc22

2005001530

ISBN-13: 978-0-393-32864-6 pbk.
ISBN-10: 0-393-32864-3 pbk.

W. W. Norton & Company, Inc., 500 Fifth Avenue, New York, N.Y. 10110
www.wwnorton.com

W. W. Norton & Company Ltd. 15 Carlisle Street, London W1D 3BS

6 7 8 9 0

CONTENTS

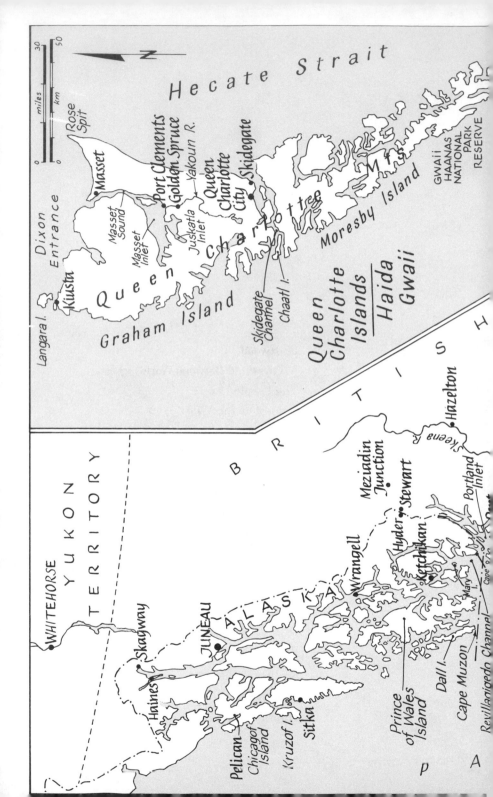

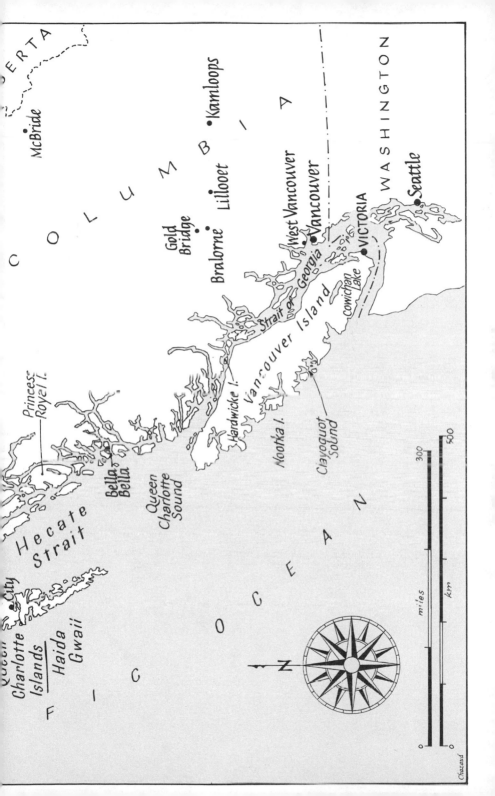

ACKNOWLEDGMENTS

THIS BOOK WOULD NOT exist if not for hundreds of acts of generosity on the part of individuals who freely shared the fruits of their hard won knowledge and experience. I wish to express my particular gratitude to the Tsiij git'anee clan, especially, Chief Donald Bell, Dorothy Bell, Lucille Bell, Robin Brown, and Leo Gagnon, for granting me permission to write about the story of the golden spruce. I am equally grateful to Cora Gray, and to those members of the Hadwin family who shared their memories and insights with me. I would also like to thank Guujaaw, the current President of the Council of the Haida Nation, for taking the time to speak with me on numerous occasions. Thanks, also, to Caroline Abrahams, Marilyn Baldwin, Perry Boyle, John Broadhead, Diane Brown, Neil Carey, Frank and Nika Collison, the late Ernie Collison, Betty Dalzell, Kiku Dhanwant and Gerry Morigeau, Bart DeFreitas and Carolyn terBorg, Tom and Astrid Greene, Paul Harris-Jones, Marina Jones, Judy Kardosh, Ian Lordon, Bruce Macdonald, Nathalie Macfarlane, Neil McKay, Jack Miller, Irene Mills, Alex Palmer, Wesley Pearson, David Phillips, Dave Rennie, Raven Rorick, Hazel

Simeon, Bill Stevens, Harry Tingley, Al and Gladys Vandal, Scott Walker, Al Wanderer, Ernie Wilson, Jennifer Wilson, and Elois Yaxley for sharing their thoughts, recollections, and excellent advice. Thanks to Archie Stocker for his video footage and photos, and to Todd Merrell, whose excellent documentary enhanced my understanding of the Haida worldview. I would also like to thank Steve Petrovcic for sharing his painstaking research, the Reverend Peter Hamel for permission to reprint his fine sermon, and the staff at the *Observer* and the *Daily News*, for their assistance with research.

Several people in the UBC Forestry Department were very helpful, including John Worrall, Suzanne Simard, and Dennis Bendickson, who fielded a random phone call and connected me to Bill Weber, whose willingness to share his time and experience was not only way beyond the call of duty but gave me a much greater appreciation for, and understanding of, loggers and the logging industry. Many thanks also to a number of people at Weyerhaeuser, especially, Erin Badesso, Bill Beese, Corey Delves, Gordon Eason, Earl Einarson, David Sheffield, and Donnie Zapp.

Corporal Gary Stroeder, Sergeants Ken Burton and Randall McPherron, and Constables Bruce Jeffrey (Ret.), John Rosario, and Blake Walkinshaw generously provided invaluable perspective on this story. Likewise, the personnel at the Coast Guard stations in Sitka, Ketchikan, Juneau, and Prince Rupert were unfailingly helpful, as were their counterparts at the Rescue Coordination Centre in Victoria. Thanks also to Grant Ainscough, Hal Beek, Paul Bernier, Pat Campbell, Don Carson, Grant Clark, Robert Fincham, Pat Fricker, Dewey Jones, Tom Illidge, Ernie Kershaw, Dale Lore, Harry Lynum, Luanne Palmer, Don Piggott, Gene Runtz, Jim Trebett, Brian Tremblay, Grant Scott, and many others, unnamed here, who have generously answered a raft of random questions.

Two superbly researched and very readable histories were crucial to my understanding of the logging and fur trade industries: Michael Williams' *Americans and Their Forests*, and James Gibson's *Otter Skins, Boston Ships and China Goods*. Jon Luoma's *The Hidden Forest* was a great introduction to the mysteries of the coastal forest. I am sorry to have missed the late poet, historian, and lifelong student of the Northwest Coast, Charles Lillard, who bears quoting here: "To read means to borrow; to create out of one's reading is paying off one's debts."

I am indebted to Steven Acheson, Robert Bringhurst, Julie Cruikshank, Robert Deal, Ian Gill, Terry Glavin, Gary Greenberg, Ben Parfitt, Roy Taylor, John Worrall, and my father, George Vaillant, for reading portions of the manuscript and commenting on their form and accuracy. John Enrico provided generous assistance with Haida translation.

Dominic Ali, David Beers, Bruce Grierson, Ruth Jones, and Jennifer Williams have been good allies. Kim and Stephen, Bree and Michael, Rikia and Cam—thank you for helping to keep the ship afloat. Thanks also to Angelika Glover and Morgen Van Vorst for their help and energizing enthusiasm. Special thanks to my editor and advocate at *The New Yorker*, Jeffrey Frank, and to the superb editors of this book: Louise Dennys at Knopf Canada and Starling Lawrence at W. W. Norton. I am particularly grateful to my agent, Stuart Krichevsky, who saw the forest for the tree.

Finally, I must acknowledge my beloved wife, Nora, a tireless editor and a true alchemist, who transformed the act of writing a first book in the midst of an emerging family into an experience I would gladly, and joyfully, repeat.

All Trees of noblest kind for sight, smell, taste;
And all amid them stood the Tree of Life,
High eminent, blooming Ambrosial Fruit
Of vegetable Gold; and next to Life
Our Death the Tree of Knowledge grew fast by,
Knowledge of good bought dear by knowing ill.

—John Milton, PARADISE LOST

THE GOLDEN SPRUCE

Driftwood

SMALL THINGS ARE hard to find in Alaska, so when a marine biologist named Scott Walker stumbled across a wrecked kayak on an uninhabited island thirty miles north of the Canadian border, he considered himself lucky. The coastal boundary where Alaska and British Columbia meet and overlap is a jagged four-way seam that joins, not just a pair of vast—and vastly different—countries, but two equally large and divergent wildernesses. To the west is the gaping expanse of the North Pacific Ocean, and to the east is the infinity of mountains that forms the heart of what some in the Northwest call Cascadia. The coastline where these worlds meet and bleed into one another is sparsely inhabited and often obscured by fog, the mountains sheared off by low-lying clouds. At sea level, it is a long and convoluted network of deep fjords, narrow channels, and rockbound islands. It is a world unto itself, separated from the rest of North America by the Coast Mountains, whose ragged peaks carry snow for most of the year. In some places their westward faces plunge into the sea so abruptly that a boat can be 50 feet from shore and still have 500 feet of water beneath her keel. The region is sporadically patrolled, being governed, for the most part, by 20-foot tides and processions of sub-Arctic storms that spiral down from the Gulf of Alaska to batter the long, tree-stubbled lip of the continent. Even on calm days, the coastline may be shrouded in a veil of mist as 2,000 miles of uninterrupted Pacific swell pummels itself to vapor against the stubborn shore.

The combination of high winds, frequent fog, and tidal surges that can run over fifteen knots makes this coast a particularly lethal one, and when boats or planes or people go missing here, they are usually gone for good. If they are found, it is often by accident a long time later, and usually in a remote location like Edge Point where Scott Walker anchored his seventeen-foot skiff on a fair June afternoon in 1997 while doing a survey of the local salmon fishery. Edge Point is not so much a beach as an alpine boulder field that, at this point in geologic time, happens to be at sea level. It lies at the southern tip of Mary Island, a low hump of forest and stone that forms one side of a rocky, tide-scoured channel called Danger Passage; the nearest land is Danger Island, and neither place was idly named.

Like much of the Northwest Coast, Edge Point is strewn with driftwood logs and whole trees that may be five feet in diameter and buried twenty deep. Burnished to silver, this mass of wood, much of which has broken loose from log booms and transport barges, lies heaped as high as polar winds and Pacific waves can possibly throw it. Even if a man-made object should make it ashore here in one piece, it won't last long after it arrives; within the course of a few tide cycles, it will be hammered to pieces between the heaving logs and the immovable boulders beneath them. In the case of a fiberglass boat —like a kayak—the destruction is usually so complete that it makes the craft hard to recognize, much less find. When a fiberglass yacht was found in a location similar to Edge Point three years after it had disappeared without issuing a distress signal, the largest surviving piece was two feet long and that was only because it had been blown up into the bushes; the rest of the sixty-foot sloop had been reduced to fragments the size of playing cards. This is why Scott Walker considered himself fortunate: he wasn't too late; parts of the kayak might still be salvageable.

The beaches here serve as a random archive of human endeavor where a mahogany door from a fishing boat, the remains of a World War II airplane, and a piece from a fallen satellite are all equally plausible finds. Each artifact carries with it a story, though the context rarely allows for a happy ending; in most cases, it is only the scavenger who benefits. Scott Walker has been scavenging things that others have lost here for more than twenty-five years, and he has acquired an informal expertise in the forensics of flotsam and jetsam. If the found object is potentially useful or sufficiently interesting, and if it is small enough to lift, the beachcomber's code will apply. Walker was abiding by this code when he happened upon the broken kayak and began tearing it apart for the stainless steel hardware.

But when Walker lifted his head from his work he noticed some things that gave him pause. Strewn further down the tide line were personal effects: a raincoat, a backpack, an ax—and it was then that it occurred to him that his prize may not have simply washed off some beach or boat dock down the coast. The more he noticed—a cookstove, a shaving kit, a life jacket—the narrower the gap between his own good luck and someone else's misfortune became. This wasn't shaping up to be a clean find. Walker deduced from the heavier objects' position lower down in the intertidal zone that the kayak had washed ashore and broken up on a low tide. The lighter objects, including large pieces of the kayak itself, had been carried further up the beach by subsequent high tides and wind, and it was these that set off alarm bells in Walker's head. Despite being wrapped around a log, the sleeping bag was still in near-perfect condition; there were no tears or stains, no fading from the salt and sun; the life jacket, too, looked fresh off the rack. Even the cookstove appeared salvageable; wedged between rocks at the water's edge, it showed only minor rusting. Winter storm season, the most effective destroyer on the coast, had only just ended, so this wreck had to be recent, thought Walker, perhaps only a couple of weeks old. Walker debated throwing the stove and sleeping bag into his skiff, but then, after considering some possible accident scenarios and recalculating the uncomfortable distance between a stranger's horror and his own delight, he decided to leave these things where they lay. Besides, he thought, they might be needed for evidence. No one would miss the stainless steel bolts, though, so he pocketed them and headed down the beach, looking for a body.

Walker never found one, and it was only through the Alaska state troopers in Ketchikan, thirty miles to the north, that he learned the story behind his chance discovery. The kayak and its owner, a Canadian timber surveyor and expert woodsman, had been missing—not for weeks, but for months. This man, it seemed, was on the run, wanted for a strange and unprecedented crime.

CHAPTER ONE

A Threshold Between Worlds

There was beauty, yes,
... but who would know until men judged it so.
　　　—Ralph Andrews, TIMBER

O N THE NORTHWEST COAST, there is no
graceful interval between the ocean and the trees, the forest simply
takes over where the tide wrack ends, erupting full-blown from the
shallow, bouldered earth. The boundary between the two is unstable,
and the sea will heave stones, logs, and even itself into the woods at
every opportunity. In return, the roots of shore pine and spruce grope
for a purchase on rocks better suited to limpets and barnacles while
densely needled boughs cast shadows over colonies of starfish and sea
anemones. The air is at once rank and loamy with the competing
smells of rotting seaweed and decaying wood. From the beach you
can see as far as height and horizon will allow, but turn inland and
you will find yourself blinking in a darkened room, pupils dilating to

fill the claustrophobic void. The trail of a person, or the thread of a story, is easily lost in such a place. Even the trees, swaddled in moss and draped in ferns, appear disguised.

A coastal forest can be an awesome place to behold: huge, holy, and eternal-feeling, like a branched and needled Notre Dame, but for a stranger it is not a particularly comfortable place to be. You can be twenty paces from a road or a beach and become totally disoriented; once inside, there is no future and no past, only the sodden, twilit now. Underfoot is a leg-breaking tangle of roots and branches and, every fifty feet or so, your way is blocked by moss-covered walls of fallen trees that may be taller than you and hundreds of feet long. These so-called nurse logs will, in turn, have colonnades of younger trees growing out of them, fifty years old and as orderly as pickets. In here, boundaries between life and death, between one species and the next, blur and blend: everything is being used as a launching pad by something else; everyone wants a piece of the sky. Down below, the undergrowth is thick, and between this and the trees, it is hard to see very far; the sound of moving water is constant, and the ground is as soft and spongy as a sofa with shot springs. You have the feeling that if you stop for too long, you will simply be grown over and absorbed by the slow and ancient riot of growth going on all around you. It can be suffocating, and the need to see the sun can become overpowering—something you could do easily if it weren't for all those trees.

FROM A SATELLITE'S-EYE VIEW, North America's coastal temperate rainforests appear as a delicate green fringe adorning the western rim of the continent. Prior to the era of industrial logging, this slender band, seldom more than 50 miles wide, stretched, virtually unbroken, from Kodiak Island in Alaska south through British Columbia, Washington, and Oregon to Mendocino County, California, a distance of more than 2,000 miles. Along these forests' entire length, a succession of mountain ranges forms a natural bulwark between the Pacific Ocean and the rest of the continent, and it is here that the storms which trundle continually across the North Pacific are stopped in their tracks. Rain clouds, functioning like airborne water bladders, burst open when they collide with the cooler air of the coastal mountains, and the results can be astonishing. During the winter of 1998 a relentless parade of low pressure systems dumped 95 feet (1,140 inches) of snow on Mount Baker, near the border of Washington State and British Columbia (a world record); at lower elevations it rained enough to float an ark.

The mild temperatures within the long, damp corridor between the Pacific Slope and the sea have created what is essentially a vast terrarium. It is an environment perfectly designed to support life on a grand scale, including the biggest freestanding creatures on earth. All of the dominant west coast species—redwood, sequoia, sugar pine, western hemlock, Douglas fir, noble fir, black cottonwood, redcedar, and Sitka spruce—are the giants of their kind. It is due in large part to these immense trees that the Northwest forests support more living tissue, by weight, than any other ecosystem, including the equatorial jungle.

The principal differences between tropical and temperate rainforests have to do with temperature and location. Whereas tropical rainforests—jungles—are found along the Equator, in the hot centers of their home continents, temperate rainforests flourish on the chilly, fogbound margins, closer to the planet's poles. These forests prefer a stable climate that is neither too hot nor too cold, and their ideal setting is a west-facing coastline backed by mountains to trap and channel large quantities of snowmelt and rain. These conditions are found in both hemispheres, but only between 40° and 60° latitude. Conifers (cone-bearing trees) in a temperate rainforest will grow continuously as long as the temperature stays above 38 degrees Fahrenheit, one reason they are able to achieve such tremendous sizes. Tree species within this climatic bandwidth vary widely, depending on where in the world they grow, but it is their relationship to the sea, even more than the trees themselves, that distinguishes these forests from their inland and equatorial counterparts.

The range of the coastal temperate rainforest—like that of most wild creatures—has been drastically reduced in a relatively short period of time. Until about a thousand years ago, temperate rainforests could be found on every continent except Africa and Antarctica. Once upon a time, the lush coastal forests of Japan were a trans-Pacific mirror of our own; mighty conifers grew there, attaining huge sizes in a climate similar to the American Northwest's. Today, with the exception of a few lone giants still standing in parks or on temple grounds, those forests are gone. The Highlands of Scotland, a place long associated with barren scapes of moorland and heather, hosted a temperate rainforest as well. So did Ireland, Iceland, and the eastern shore of the Black Sea. While the North Sea coast of Norway retains vestigial traces of its original rainforest, Chile, Tasmania, and New Zealand's South Island are the only places left with forests whose flora, feel, and character remotely resemble those of the Pacific Northwest, which hosts the largest such forests in the world.

Like Tolkien's Ents, the trees of the Northwest have been marching up and down the coast for eons, fleeing southward with each ice age and reclaiming lost territory as the glaciers recede. The current rebound is still under way, with the result that Sitka spruce are advancing northward into Alaska at a rate of about one kilometer every century. Western red cedar, the tree from which Northwest Coast tribes derived virtually all of their building material, has been in existence in its present range for only four or five thousand years. Thus, while individual species may be ancient and the trees may qualify as "old growth," the forests that contain them are mere children by geologic standards, and even by our own. By the time the first of these trees matured, human beings had been living in North America for at least 5,000 years.

Until recently, North America's coastal rainforest was so poorly understood that even within the logging industry it was referred to as a "biological desert." While the process of cataloguing and understanding the creatures that share the forest with these trees is still in its infancy, it is known that the forest floor, as well as the canopy above, is almost literally seething with life. It has been estimated that a square meter of temperate forest soil can contain as many as two million creatures representing a thousand species. Andy Moldenke, an entomologist at Oregon State University, calculated what might be found within the area of an average-sized shoe; he determined that a single footstep in one of Oregon's coastal forests is taken on the backs of sixteen thousand invertebrates.

Most of this activity occurs unseen, but on some level it can be felt. The atmosphere in an old-growth coastal rainforest borders on the amniotic; still and close, sound moves differently in here, and the air moves hardly at all. Because of the forest's proximity to the coast, the sea and many of its inhabitants are a strong presence *within* the forest itself. Thriving on the instability of high-latitude ocean weather and its attendant smorgasbord of nutrients, the entire ecosystem comprises a hydroponic matrix in which behaviors and boundaries we take for granted are crossed and, in some cases, reversed. Depending on tides and rainfall, salmon and trout, returning from their transoceanic odysseys to their home rivers, can be found stranded in the branches of trees while ancient murrelets, an elusive seabird that "flies" underwater, will nest beneath their roots. Ten stories above the forest floor, their close relatives, the marbled murrelets, launch their own subaquatic feeding missions from moss-covered nesting platforms that may be centuries old. Reaching speeds of 100 miles an hour, they hurtle to and fro   forest to sea and back again—like bumblebees on speed. Moving at one one-hundredth that velocity, ocean-fed bears— some of them as white as a bald eagle's head—swim from island to island where they cruise the high tide lines, their footprints overlapping with those of deer, otter, marten, and wolf. Meanwhile, seals will pursue saltwater fish deep into the forest, hauling out to rest themselves next to a tree that might have been a bear's den the previous winter. In here, the patient observer will find that trees are fed by salmon, eagles can swim, and killer whales will heave themselves into the graveled shallows and stare you in the eye.

The Northwest Coast Indians spent most of their lives within a hundred yards of this heavily trafficked threshold between worlds. Living in such a liminal environment, it is hardly surprising that their artworks, dances, and stories focus so heavily on convergence and transformation. Nowhere else on the coast is the profound interdependence between the forest, the sea, and their shared inhabitants more dramatically represented than on the Queen Charlotte Islands. Named after the ship of an eighteenth-century British trader, the islands are the historic territory of the Haida people, who live there to this day and who call their home Haida Gwaii. On maps, the wing-shaped archipelago, comprising more than one hundred and fifty islands and islets, appears to have broken loose from the continent and gone to sea, leaving behind a noticeable hole in the snugly fitting puzzle of inlets and islands that defines the Northwest Coast. The closest land is Alaska's Prince of Wales Island, a forty-mile sea journey to the north. British Columbia, of which the Queen Charlottes are the most far-flung part, lies fifty miles to the east. To the south and west is open ocean, but it is not a gradual sloping into the depths of the Pacific; it is a near-vertical plunge. The 180-mile-long archipelago is perched on the outer edge of the continental shelf, which here takes the form of a 9,000-foot submarine cliff. Along the islands' storm-scarred west coast, this sudden shift in sea depth generates waves big enough to deposit drift logs atop 100-foot cliffs, and baffling currents that can cause tides to flow, not in two directions but four. Following the archipelago's seaward contours, nearly two miles down, is the Queen Charlotte Fault, where the northbound Pacific Plate and the southbound American Plate grind past each other with excruciating slowness and devastating force. The epicenter of one of the most violent earthquakes ever recorded anywhere on the West Coast (8.1 on the Richter scale) was located here.

If the Hawaiian Islands had risen from the sea 3,000 miles farther north and east, the Queen Charlottes are what they might have looked like. The islands are, in effect, a moated rainforest clinging to the shoulders of snowcapped mountains, and they are not an easy place to get to: Vitus Bering had explored the Alaskan coast and

Captain Cook had made landfall in Australia before Europeans ever set foot in the Charlottes. Even now, the journey from Seattle or Vancouver, via car and ferry, takes three days. Europeans have acknowledged these islands' mystical and revelatory qualities to the point that even loggers and land-use planners employ the adjective "magic" to describe them. Perry Boyle, a veteran tugboat operator from Prince Rupert, on the adjacent mainland, may have summed it up best when he said, "Everything is mythical over there." This land lying "west of west" represents a concentration of what one might call geographic essence, as if the nature and spirit of a much larger region were compressed into a space too small for it to plausibly hold. Greenhouses, libraries, and museums can simulate this effect, and Jerusalem is an example of the real thing, as are the Aran Islands, Yosemite National Park, and Delphi. Lower Manhattan is a modern urban version, and the cathedral at Chartres is an ecclesiastic one. For many West Coast Canadians and others familiar with this part of the world, the Charlottes represent a kind of "soul home," a wild native Eden; even if they haven't been there, it is a place whose existence they find at once stimulating and reassuring. The islands provide a link to how things were before the arrival of Europeans as well as a glimpse of a possible future.

A vivid example of how these islands seem to be a concentration of something much larger was reported by a turn-of-the-century American hunter and naturalist named Charles Sheldon. Sheldon traveled extensively across the West, including the Northwest Territories and Alaska, and he wrote several books about his adventures that have become classics of the genre. In the fall of 1906, he was lured to the Charlottes by rumors of an exceptionally rare subspecies of caribou, known only to the islands. While searching for a worthy specimen, he spent an astonishingly wet month on foot, exploring the north end of Graham Island, the largest in the archipelago. His quest led him through deep forests, up rivers, and across treeless swamp barrens where he noted a bizarre phenomenon: "One conspicuous feature of the atmospheric effect in that locality was an optical delusion exactly the reverse of that common on our Western plains of the United States. Objects appeared very distant when they were really very near, and it required a long time to become accustomed to the short spaces actually traversed when to the vision they appeared so long."

There is no doubt that these islands have a powerful effect on people and, as with Sheldon's observation, the light may have a lot to do with it, perhaps because it is meted out so grudgingly. The Queen Charlotte Islands are among the rainiest places in North America; they occupy a region known to ecologists as the Very Wet Hypermarine Subzone where the total hours of cloud cover amount to more than 250 days per year. When the sun does shine it is often through a prism of water particles, and for this reason rainbows are a common occurrence here. Far rarer but documented nonetheless are lunar rainbows; they appear as luminous ghost arches caused by a rising or setting moon shining under rain clouds. But there is more to it than water and light; the life force out here is extraordinarily strong in a literal, biological sense. Twenty-three species of whale live in or pass through the region's waters, and the islands themselves are home to one of the continent's highest concentration of (resident) bald eagles. Burnaby Narrows, a slender tidal channel in the middle of the archipelago, contains one of the highest concentrations of sea life per square meter of any place on earth. Meanwhile, the islands' saw-toothed west coast produces mussels the size of dress shoes.

EVER SINCE THE END of the last ice age, the Queen Charlottes have been on their own, and responsibility for this lies solely with Hecate (HECK-et) Strait. Within a space of only fifty miles the sea depth around the islands changes from 10,000 feet to less than 200. This relatively rapid decrease, combined with exposure to the full brunt of severe polar storms and huge Pacific rollers, can cause the strait to explode from a flat calm to 60-foot waves in a couple of hours. The broad, shallow channel—barely a hundred feet deep in some places—was named after the British paddle-wheel sloop H.M.S. *Hecate*. Armed with heavy guns, the vessel was brought up to the Charlottes in 1861, both to survey the surrounding waters and to ensure that recently arrived copper miners wouldn't be attacked by the Haida Indians. Naming geographical features after one's ship was a common practice in the eighteenth and nineteenth centuries, but few of these names fit their subjects as well as the *Hecate*'s did. Hecate is a Greek goddess of sorcery and witchcraft often associated with fishermen and the land of the dead. According to the *Oxford Dictionary of Classical Myth and Religion*, she is "intrinsically ambivalent and polymorphous. She straddles conventional boundaries and eludes definition." She has been depicted with man-eating dogs for feet and is known as a source of abundance of all kinds, including storms. "She's a black-hearted bitch," said one veteran fisherman of the strait. "Sometimes I think she just wants to keep the Charlottes for herself." Even now, heavy seas routinely delay the 552-foot-long passenger ferry that connects the islands to the rest of the continent. The seven-hour journey can be so rough that trucks must be chained to the deck like container vans on a transoceanic voyage.

GRAHAM AND MORESBY ISLANDS form the spine of the southward-tapering archipelago, and even though Moresby is only five miles wide in some places, it surges skyward in a wedge of sharp new mountains nearly a mile high. Hundreds of waterfalls and dozens of creeks and rivers pour out of these mountains on all of the larger islands, among them, the Yakoun (ya-KOON). The Yakoun River rises in the Queen Charlotte Mountains at the south end of Graham Island, and it gathers in Yakoun Lake before heading northward toward Masset Inlet and the sea. As the archipelago's longest river and the source of its largest trout and salmon runs, the Yakoun represents the aorta in the greater body of the islands. The low-lying alluvial valley through which it flows is well known for its massive old-growth timber, particularly its high density of knot-free and

straight-grained Sitka spruce. A valley bottom like this is what commercial loggers—who would not arrive in the Charlottes until the
twentieth century—would come to know as a "spruce flat." Here, the
soil is deeper and richer than on the mountainsides and this, combined with the Queen Charlottes' mild climate and an annual deluge
of rain amounting to as much as eighteen *feet*, creates ideal growing
conditions not just for Sitka spruce but for its common neighbors,
western hemlock and western redcedar. Hemlock and spruce in particular are commonly nurtured by nurse logs; these dead, rotting
humus-rich trees provide a ready feast for a seedling, much the way
the fruit of an apple feeds its seeds. As a nurse log is consumed by the
young forest around it (a process that can take hundreds of years), the
younger trees may be left standing well off the ground on stiltlike
roots. Over time, the gaps fill in, but it is not uncommon to find a 400-
year-old Sitka spruce with a tunnel beneath it large enough to crawl
through.

Of all the West Coast conifers, the Sitka spruce seems the most
naturally suited to the maritime environment. Its long, narrow geographic distribution mirrors that of the Pacific rainforest, and the
species shows a preference for planting itself in the teeth of the gale.
Sitka spruce have a high tolerance for salt spray and they often serve
as the first line of defense between the sea and the forest; their great
size and strength breaks storm-driven winds that can lay waste to
lesser species. Sitka spruce is the world's largest and longest-lived
species of spruce; it can live for more than 800 years and grow to
heights exceeding 300 feet, which is tall even for a redwood. Despite
the colossal end result, their beginnings are almost unimaginably
humble: a single Sitka spruce seed weighs only 1/13,000 of an
ounce, and yet it contains all the information needed to produce a
tree that can weigh more than 300 tons—about as much as three
blue whales. While the species is common up and down the coast,
these "mega spruce" grow in only a handful of places, and one of
them is the Yakoun Valley.

During one autumn around 1700, on the west bank of the Yakoun

River, a random Sitka spruce cone opened and let a seed like no other drift to earth. It was one among hundreds of seeds that fell that year from one among thousands of cones produced by one among the tens of millions of Sitka spruce trees growing on the Northwest Coast. Its parent tree may well have been scattering seeds since the time of the Vikings. Were it not for the fact that an individual spruce seed's chances of survival are comparable to those of a human sperm, every parent tree would be a forest unto itself. As it is, despite as much as 750 years of fertility, a typical Sitka spruce may produce only a dozen offspring that survive to maturity. That the seed in question would be one of them mystifies people to this day.

Shaped like a teardrop and about the size of a grain of sand, the seed would have appeared identical to all the others that had been peppering the forest floor for millennia. Of its cone mates that landed on the heavy moss carpeting much of the forest, only one in a hundred would germinate. Those lucky enough to land on a nurse log would fare better, but even then the chances were one in three that they would be killed by fungi within the month. Somehow this anonymous seed with its strange message beat these abysmal odds and managed to take root. The tiny sprig would have been easy to miss in the crowded nursery of the forest floor, surrounded as it was by thousands of other aspiring trees—not just Sitka spruce but hemlock, redcedar, yellow-cedar, and the occasional yew. At this stage, it would have been dwarfed by everyone, even habitual shadow dwellers like little hands liverwort, lover's moss, black lily, sword fern, and devil's club, not to mention the dense thickets of salal that can grow a dozen feet high and require a machete to penetrate.

To look at this seedling—if one could see it at all—and believe that it had every intention of growing into one of the towering columns that blot out so much of the northwestern sky, would have seemed far-fetched at best. In its first year, the infant tree would have been about two inches tall and sporting a half dozen or so pale green needles. It would have been appealing in the same abstract way that baby snapping turtles are, its alien appearance transcended by the universal indicators of wild babyhood: utter helplessness and primordial determination in equal measure. Despite its bristling ruff and a stem as straight as a sunbeam, the seedling was still as vulnerable as a frog's egg; a falling branch, the footstep of a human or an animal—any number of random occurrences—could have finished it there and then. Down there in the damp darkness of the understory, the sapling's wonderful flaw was a well-kept secret. With each passing year, it dug its roots deeper into the riverbank, strengthening its grip on life and on the land. In spite of the odds, it became one of a handful of young trees that would survive to shoulder their way into the sunlight, competing with giants a dozen feet wide and hundreds of feet tall. In the end, it would be the sun that exposed this tree's secret for all to see, and by the middle of the 1700s it would have been abundantly clear that something extraordinary was growing on the banks of the Yakoun. It was a creature that seemed more at home in a myth or a fairy tale: a spruce tree with golden needles.

Unless a tree is particularly large, or unusually shaped, it will not stand out as an individual, and unless it is isolated from its mates, it will seldom announce itself from a distance. But despite being embedded in a forest of similarly large trees, the tree that came to be known as the golden spruce was an exception on both counts. From the ground, its startling color stopped people dead in their tracks; from the air, it stood out like a beacon and was visible from miles away. Like much of the surrounding landscape, the tree was incorporated into the Haida Indians' vast repertoire of stories, but as far as anyone knows, it is the only tree, in what was then an infinity of trees, ever to be given a name by the Haida people. They called it K'iid K'iyaas: Elder Spruce Tree. According to legend, it was a human being who had been transformed.

Although it was well known to those who lived around the Yakoun Valley, the golden spruce wasn't discovered by scientists until well into the twentieth century. By then it was more than 200 years old and all but impossible to miss. When the Scottish timber surveyor and baronet Sir Windham Anstruther stumbled across the tree in

1924, he was dumbfounded. "I didn't even make an ax mark on it," he told one reporter before he died, "being, I suppose, a bit overcome by its strangeness in a forest of green." For years afterward, no one knew quite what to make of Sir Windham's arboreal unicorn. Some suggested it might be a new species, native to the archipelago; others supposed the tree had been hit by lightning, or was simply in the process of dying. As it turned out, the tree was alive and well; it was just fantastically rare. So rare, in fact, that it warranted its own scientific name: *Picea sitchensis* 'Aurea.' *Picea sitchensis* is the Latin name for the Sitka spruce, and *Aurea* is Latin for "golden" or "gleaming like gold," but it can also mean "beautiful" or "splendid." Sixteen stories tall and more than twenty feet around, the golden spruce was unique in the botanical world.

CHAPTER TWO

The People

The island was nothing but saltwater, they say. Raven flew around.
He looked for a place to land in the water. By and by, he flew to a reef . . .
to sit on it. But the great mass of supernatural beings had their necks
resting on one another, like sea cucumbers. The weak supernatural beings
floated out from it sleeping, every which way, this way and that way
It was both light and dark, they say.

    —from "Raven Who Kept Walking," a Haida creation story

THE HAIDA VILLAGE of Old Masset hugs the
beach on the eastern shore of Masset Sound at the upper end of
Graham Island. The sound is a broad channel that winds through
dense forest and swampland, and in the course of cutting the island
almost in half, it links the Yakoun River to the sea. Big, determined
tides push and pull along its serpentine length, and are felt as far
upstream as the golden spruce, more than thirty miles to the south.
Just past Old Masset's graveyard, this brackish two-way river makes a

final dogleg turn around a spit of sand before emptying into the broad gap between Graham and Prince of Wales Islands, a nasty stretch of water called Dixon Entrance. Fully exposed to the Pacific, it is one of several gateways for the sudden tempests that plague Hecate Strait. Even on the calmest days, the sea rolls by in long hillocks, the lumbering, whale-backed memories of storms that once wracked Hokkaido, Kamchatka, or the Aleutians.

Along the beachfront at Old Masset, monumental poles—the carved spines of trees—stand vigil. Many have been raised to honor the dead, but at the north end of the village, in front of a prominent chief's house, there is one with a different purpose. The chief himself is a master carver, and his house is an imposing structure of broad cedar planks and heavy beveled beams. It stands apart from the other village houses, which have been built in orderly rows, closely following the contours of the shore. Most of the houses and all of the poles are oriented toward Masset Sound, but his pole and the ferocious creatures that compose it are angled away, toward the open sea. The pole is around forty feet tall and four feet through at the base. Its lower section is carved in the shape of an enormous grizzly bear, and cradled in its forepaws is a dugout canoe. To get an idea of the weather that blows through here on a regular basis, one need only look inside this canoe; though it is ten feet off the ground, it must be emptied periodically of windborne sand and seaweed. There are other animals higher up on the pole, and there is an eagle on the top—an indicator of the chief's lineage—but it is the bear and its carefully held canoe that grab the eye and hold it. There is something strangely familiar about them, but it takes a moment to realize what it is.

On the opposite side of the continent, in another small fishing town, there is a statue of a human being named Mary who can be seen holding a boat of her own. What is difficult to determine with certainty about either the spirit bear in Old Masset or the spirit woman in Gloucester is whether they are truly protecting these vessels, or simply preparing to offer them up. In any case, generations of Gloucester fishermen and their families have knelt before Our Lady of Good Voyage and prayed for the safe passage of their ships, their loved ones, and themselves. And on a soft spring afternoon in 2003, in the parallel universe of Old Masset, a similar ritual is taking place at the foot of the chief's pole. If you had been there that day, and you happened to close your eyes, relying solely on your remaining senses, time would have slipped out from under you. You would have found yourself grasping at centuries

In a pit nearby, a driftwood fire is burning, and a cedar plank arrayed with slabs of carefully seasoned salmon and halibut has been laid upon the flames. But none of the people who stand and sing around the fire has any intention of eating this rough feast; these delicacies are not for them. The smoke moves from quarter to quarter in the testy wind like a broken compass needle as the fish incinerates, its essence corkscrewing into the cloud-streaked sky on its way to feed Skilay. Skilay was the spokesman for the golden spruce, and now he is dead. Today the people have gathered by the hundreds to fill the dark hole he has left behind.

IN 1859, WILLIAM DOWNIE, a successful gold prospector (for whom Downieville, California, is named), traveled to British Columbia, where he worked both as a prospector and as an explorer for the British colonial governor. In the course of his travels Downie visited the Queen Charlottes, where some large gold strikes, including a single 21-ounce nugget, had been found. In his report to the governor, Downie wrote that he had found the Haida were "first-class prospectors, and know all about gold mining." But he was even more impressed with their seamanship: "They are the best boatmen I have ever met, and in saying this I refer to both sexes. They have, indeed, an amphibious-like nature, for they seem to be as much at home in the water as they are ashore, and for feats of diving and swimming their equals are not easily found."

In 1873 James Swan, an American writer, judge, historian, customs collector, and promoter of the Northwest, visited the Queen Charlottes on behalf of the Smithsonian Institution. While there, he reported seeing canoes that were "very large and capable of carrying one hundred persons with all their equipments for a long voyage [*sic*]."

Native peoples throughout the Americas have used dugout canoes (hewn from a single log) for a variety of purposes ranging from off-shore whaling expeditions to warfare and the transport of people and trade goods. Dugouts were used widely on both coasts of North America, but it is the Indians of the Northwest who are credited with building the biggest canoes made by anyone anywhere in the world; some approached 100 feet in length. Once a canoe-worthy tree had been selected, it would be felled with stone axes and fire; carvers would rough it out on the spot and then drag it—sometimes for miles—through the woods to the carver's village for finishing. Failed attempts can still be found scattered throughout the forest. Traveling in these giant cedar canoes, the Haida would regularly paddle their home into, and out of, existence. With each collective paddle stroke they would have seen their islands sinking steadily into the sea while distant snow-covered peaks scrolled up before them like a new planet. Few people alive today have any notion of how it might feel to pull worlds up from beyond the horizon by faith and muscle alone.

Like all the coastal Indians, from Northern California to southeast Alaska, virtually everything the Haida made began as a tree. Their hats and baskets were woven from spruce roots, and just about everything else, including much of their clothing, came from the bark and wood of the red cedar; tall, straight-grained, and easy to work, it encourages construction on a massive scale. Their houses are the size of small airplane hangars; their carved poles can be as long as their canoes. Launching from their remote marine base, the Haida raided and traded up and down the coast as well as far into the interior by river. They suffered casualties but rarely retribution because few of the coastal tribes had the skill or audacity to pursue them across Hecate Strait. While some of these journeys were dedicated to peaceful commerce, many expeditions—even those to neighboring villages—were devoted to raiding and slave-taking. By 1850, the tribe had become legendary for a ferocity, mobility, and naval daring comparable to the Vikings. There has been a great deal of speculation about how far the Haida traveled, but it has since been demonstrated that a nineteenth-century dugout canoe can make the journey from B.C. to Hawai'i. (Based on existing trade routes and maritime technology it would have been possible—in theory—for a Greek to get to California as early as 400 A.D.)

Several adjacent mainland tribes, the Tlingit and Tsimshian in particular, had reputations as fierce as the Haida, though with more territory, trading opportunities, and enemies at their immediate disposal, they tended not to travel as far afield. Despite their mutual hostility, all the Northwest Coast tribes share strong cultural ties. They traveled by canoe, carved poles, and had similar tribe and clan structures; in some cases they intermarried, and they all attached a high value to wealth and status which found its purest expression in the potlatch ceremony. A potlatch can serve many purposes, from celebrating the construction of a house or demonstrating an individual's worthiness to lead, to saving face or making amends for an injury— social or physical—that has been suffered upon a member of another family or clan. They are also held to acknowledge the passing of a remarkable person. No matter what the purpose, the host provides food and gifts for all who attend, thus obligating them to be witnesses to whatever has transpired. The tribes of the Northwest Coast are the only ones on the continent who had so many possessions, and such effective means of transporting them that they could store their belongings in heavy wooden chests, some of which can hold a person with ease.

In addition to being master mariners, the Haida were sea hunters who pursued shark, seal, sea lion, halibut, and occasionally whales. But it hardly seemed necessary; shellfish were so abundant, and the runs of salmon, herring, and pilchard, among others, so vast and easily harvested, that the Haida's environment could be described as a kind of subaqueous buffet enlivened by revolving seasonal specials. Whatever they lacked on the islands they could trade or fight for on the mainland. Even today, island bays will turn white with herring milt (semen), and flocks of seagulls a mile wide and many miles long can be seen pursuing eulachon* up the Skeena River, across Hecate Strait. It was because of this bounty that the Northwest Coast had one of the densest nonagricultural populations on earth. With food so plentiful and the climate so moderate, the Haida, like their tropical counterparts, had an enormous amount of free time to feast, fight, tell stories, make monumental art, and build gigantic dugout canoes—in short, to develop a highly complex culture. It has been estimated that as much as 40 percent of the region's inhabitants were slaves.

The Haida's masks, "totem" poles, bighouses, and canoes repre-sent a high point in North American art and craftsmanship. Without knowing who made them, most people would recognize their art-works, which have become international symbols of North American native culture. A Haida canoe, fifty-four feet in length, is on perma-nent display at the Canadian Museum of Civilization in Ottawa. An even larger canoe—sixty-three-feet long and elaborately decorated—is the centerpiece for the Northwest Indians exhibit at the American Museum of Natural History in New York.†

Many traces of this early legacy remain. Dotted throughout the islands that constitute the Haidas' historical domain are abandoned villages where one can still see the same redcedar poles that greeted and alarmed the archipelago's first European visitors. Nowhere else on the coast—or in the world—do so many old poles survive in their original beachfront locations. Cedar is exceptionally durable, but out here, a typical pole lives only about as long as a human being before it falls over and is consumed by the forest. These poles are the Easter Island and Angkor Wat of the Pacific Northwest, but where the lat-ter could last indefinitely, the Haida poles' woodenness is their death sentence; in all likelihood any poles still in situ will revert to nature

---

*An oily, finger-sized fish and highly valued trade item that can be eaten, rendered for its oil, or stood on end and lit like a candle.
†While attributed to the Haida, this canoe was probably made by the Heiltsuk people who live on the central mainland coast of B.C.

in our lifetimes (in accordance with the Haida's wishes). The village of Nan Sdins (NIN-stints), at the southern tip of the archipelago, is the most famous and best preserved of these places, and it has been classified as a UNESCO Heritage Site. Due to the village's sheltered location and recent conservation measures, more than two dozen poles still stand here despite being well over a hundred years old.

Half of these poles are visibly fire-scarred because, once the village had been abandoned at the end of the nineteenth century, members of one of the coastal tribes who had been raided repeatedly by Nan Sdins warriors crossed Hecate Strait and set fire to the village in an act of long-deferred revenge. Today, it is still possible to see what fire, anthropologists, and time have left behind. Bleached like bones, the fixed and staring eyes of eagle, raven, killer whale, frog, bear, and beaver—heraldic crests and spiritual allies of the former inhabitants—gaze back from the intentional forest of poles. Carved from single trees, the creatures are stacked upon one another dozens of feet high, merging together as if sample specimens of the local fauna humans included—had been stuffed, one after the other, into giant test tubes and petrified. Their deftly carved features are exaggerated and intimidating: tongues loll, nostrils flare, teeth are bared, but now these expressions seem more the effects of rigor mortis than of the vigorous ferocity of life; this is a place of ghosts. It seems appropriate, then, that almost all that remain are mortuary poles, once topped with bentwood boxes filled with the remains of wealthy people. One struggles to imagine the lives lived out there: the ambitious sculptors thrilled to be carving with European iron rather than Tsimshian beaver teeth; the barn-sized longhouses made of cedar planks and posts; the lavish potlatches in which chiefs and nobles gained status by demonstrating how much they could afford to give away.

IT IS A MEMORIAL POTLATCH for Skilay that has brought so many people to Old Masset. Skilay was one of the most powerful members, not just of his clan but of the entire Haida Nation. He was not a chief, but he occupied a position that was equally admired and, in day-to-day life, even more practically important. A talented fisherman, carver, and singer, and dedicated politician and activist, he was one who could transcend boundaries; when everyone else was too angry or too discouraged to talk, he could get them laughing again. Skilay was known as the Steersman; he was one who made sure the canoe—the Haidas' ship of state—was going in the right direction. For many of the assembled, Skilay is, warts and all, a living embodiment of what it means to be Haida—in other words, a human being.

The word "Haida" simply means "people," which is really just another word for "us." In fact, throughout the world, the names used by most indigenous peoples to describe themselves translate to this, the implication being that "We are Us: the People—and the rest of you are not." The Haida call their island home Haida Gwaii, which means, literally, "Place (Islands) of the People," but there is an older name, and it translates, roughly, to "Islands Coming Out of (Supernatural) Concealment." In this sense, the islands represent a sort of existential intertidal zone—not just between the forest and the sea but between the surface and spirit worlds. Haida Gwaii is the most remote archipelago anywhere on the West Coast, and there is no other North American tribe whose ancestral home is located further offshore, or whose territorial boundaries are so clearly and unambiguously delineated. It is generally believed that parts of the islands were *refugia*, places left untouched by the great ice sheets which covered so much of North America during the last ice age. As a result, these islands are sometimes referred to as the "Canadian Galápagos," and in many ways they are a world apart, hosting numerous species and subspecies that occur nowhere else. The Haida language, too, is what linguists refer to as an "isolate," unrelated to that of any other West Coast tribe.

Like the vast ocean and the fitful weather that surround them, virtually everything in the Haidas' world is capable of changing form and function as whim or circumstance dictate. Thus, a rock is never just a rock, and a crab is always more than a crab. Mountains can take the form of killer whales, and a canoe can open its mouth and tear out the throat of a grizzly bear. Virtually every rock, reef, island, and inlet in the archipelago has some supernatural association, just as prominent geographical features in Australia's Outback do for Aborigines, and those in the Holy Land do for Muslims, Christians, and Jews. The golden spruce is woven into this web of shape-shifting, interconnected meaning as well, and many representatives from these multiple dimensions have been summoned to Old Masset in order to honor Skilay.

Skilay was an Eagle, one of two primary Haida tribal affiliations, or moieties (the other being Raven), and beneath these heraldic umbrellas are dozens of clans. Moiety and clan affiliation are inherited from one's mother and each is represented by a crest. While most of these take the form of birds, animals, sea creatures, or humans, far more abstract symbols such as rainbows, clouds, and even avalanches are used as well. As a result of intermarriage, most families possess several crests, and when it comes to sheer, hyphenated complexity, the peoples of the Northwest Coast make European noble families look like amateurs. Anthropologists have compared their kinship systems to higher math. Skilay's primary clan affiliation was Tsiij git'anee [cheets-GIT-nay] (Tsiij Island Eagle People), the same clan whose historic territory includes the northern reaches of the Yakoun River which encompass the land around the golden spruce. In addition to his many other roles, Skilay represented K'iid K'iyaas, a being he loved and whom he and his clan were obliged to protect. But Skilay was killed in his prime, and now, nearly two years after his death, the family has everything in place for his memorial potlatch. They have amassed the capital, bought and made the gifts, sent many hundreds of invitations, prepared the food, and paid for the carving of a carefully chosen forty-foot-tall cedar pole. Properly executed, all of these things will ensure that suitable honor is done, not just to their beloved Skilay but to his family, his clan, his moiety and his tribe.

Skilay was a big man who loved to cook and to eat; he was a good provider and generous to a fault—so generous that he would take in people no one else wanted—even those outside the tribe. Skilay adopted an Anglo boy called Bone—short for Bonehead—and he gave him a family, a tribe, and a life worth living. Bone is big and bald and he will carry the heavy soup pots at his adoptive father's memorial feasts; he will clean the great hall each morning after the hundreds of guests have gone to bed at sunrise, laden with gifts. On the second night of feasting, he will be given a proper Haida name and then he will surprise everyone with his eloquence.

Like all potlatches, this one has been carefully choreographed, and the food ritual being performed beneath the chief's pole is just one part of an elaborate process that will take days to complete. Even as the salmon and halibut turn to ash and vapor, the chief himself puts the finishing touches on Skilay's memorial pole, which is topped with a hummingbird, the Tsiij git'anee clan crest. Later in the day, the pole will be raised beside his house, a difficult and dangerous task that will take literally hundreds of people to accomplish. Guests have prepared for months and traveled for days to be here; they have come from the north and the south and the mainland. Most of the senior guests wear spruce root hats. Woven tight enough to shed the rain and broad enough to block the sun, they are painted in highly refined and stylized designs of black and red; some of the brims are hung with ermine pelts and miniature canoe paddles which play about the wearer's face. On their wrists, the wealthier women wear heavy cuffs of gold and silver that look like treasures from a pharaoh's tomb. To those within the tribe, it is clear at a glance who the esteemed artisans are as well as the nature of their connection to the wearer. Artist and patron alike are, in most cases, nobility, and from these ornaments alone can be deduced more about the wearer's lineage and income—her place in the community, the tribe, the world—than any modern credit check or social security number ever could supply.

The chiefs and their powerful wives have come wrapped in the skins of bears, in shawls of mountain goat fur, and in capes of leather and melton wool trimmed with ermine skins and abalone buttons; some carry heavy talking sticks as tall as a man. Like the ladies' bracelets, all this regalia is decorated with tribal and family crests: raven, eagle, frog, bear, and berry-woman-in-the moon, among many others, and all are more than they seem. The boundary between simply donning a cape and assuming the mantle of another being is a fine one. Like the poles that stand in front of the village's more important homes and buildings, the assembled hats, capes, bracelets, and pendants represent a sort of cosmic social register. They link together, through their woven, painted, and deeply scriven hands, paws, claws, talons, fins, and flippers, everyone from immediate family to the remotest spiritual allies and animal ancestors. It is for this reason that after a dance performed to honor people of the Eagle moiety, a hall may fill with the shrill, dry, and unmistakable sound of eagle whistles, as if those gathered there had been temporarily inhabited by birds. One need only imagine this animal energy redirected in armed and painted anger to get a sense of the bowel loosening terror felt by foreign adversaries.

IT IS MIDDAY and the offering of salmon and halibut has been received; scent and substance have blown away in the shifting wind that hustles now, southbound down the sound. The feast of flesh is followed by an offering of the spirit, and it is delivered in a box made from a single plank of cedar that has been notched, steamed, and bent into a perfect cube. Ordinarily these bentwood boxes are elaborately decorated, but this one has been left blank, painted black. It is a box in mourning, a box that contains something best left unlabeled. Inside it is a mask that took weeks to carve and that, if sold, would go for thousands of dollars. But this is not the kind of mask that can be bought or hung on a wall; it cannot be reused in any way. This is Skilay's spirit mask; it can be danced only once, and its time was last night. The dancer who danced it with its sightless eyes fixed in a pale moon face was led about the crowded hall by another dancer shaking a rattle. Drums were pulsing from different corners of the great room, moving through the bursting crowd and coalescing with stomping feet into a floor-shaking tumult that rumbled like big stones in the surf. Outside, it was blowing hard and raining. Huge, heavy-browed ravens stalled against the driving wind and hung there, motionless by the roof peak, and then, with an imperceptible tilt of a blue-black wing tip, they would disappear, as if jerked away by an unseen string. Inside, the singers' voices rose into the air in hackle-raising frequencies, resonating with the overtone of grief that permeated the room as more dancers in masks representing Frog, Eagle, and other spirit beings from beyond welcomed Skilay home. Skilay's body was dead and buried, but this was really it: his spirit was leaving town, and there was scarcely a dry eye in the house.

As the flames rise around the sealed black box, the people continue to sing. For a long time the box seems to sit in the fire as if it were comfortable there, but in time some cracks begin to show. As the box becomes fully engaged, a bag is passed, and one by one the people break out of the circle to sprinkle tobacco on the flames and reveal their private thoughts to the man they loved and admired. As if on cue, a bald eagle alights in the top of a nearby spruce, and for a moment she and the carved eagle atop the adjacent pole neatly bracket the chief's house. But nothing here is new to her, and after a time she leans forward and, with a few downward thrusts of heavy wings as broad as a man is tall, she finds a draft, locks into it, and glides away. Shortly afterward, a strange thing happens: all at once, the top and sides of the box spontaneously lift off and fall to the side. It is hard to explain this in any structural or thermodynamic way, but it happens suddenly and, for a brief moment, the mask stares out from the pit, engulfed and yet untouched by the fire. The geisha-white face shines around the scarlet lips as flames burst from the eyes, mouth, and nostrils. When, at last, the heat becomes too much and the finely carved cheeks split beneath each eye, they do so simultaneously, along the grain, and it looks to some as if the mask is weeping molten tears. What is the carver feeling at this moment, before the chin and forehead give way and his labors crumble into glowing embers? What is happening in the hearts and bellies of Skilay's children and the somber chief as the dim shadow of a grizzly bear holding an empty canoe clocks slowly across the ground?

BY MIDAFTERNOON SKILAY'S POLE is finished and, with the paint still wet, the men gather to move it to his house. It is shockingly heavy—frighteningly so: the pole is a dozen feet around and weighs six-and-a-half tons. Once again, the hole left by Skilay yawns open. He has always been the one to supervise the pole raisings. Will the pole still rise without him? Will this be the time someone gets killed? The first steps are awkward: a leg is almost crushed; decisions are made, not by an experienced leader, but by the group, in the same way that a school of fish decides to turn in a particular direction. Different leaders emerge and then recede and, in this way, with Eagles on one side and Ravens on the other, the pole finds its way to the grave-sized hole next to Skilay's house. But the hardest part is yet to come; maneuvering this giant statue into a standing position will be brute proof of the people's devotion—the most difficult thing anyone there will do for a long time. One reason Skilay's pole is so heavy is that it is a solid cylinder, not a hollowed-out half pole like so many others. And unlike those lighter poles, this one is deeply carved, not only from top to bottom but all the way around. As with the weight and complexity of a gold bracelet, these details are all indicators of Skilay's stature, and of his family's wealth. The fact that his pole was carved by one of the best living carvers on the coast, is further evidence of Skilay's position in the tribe.

Ten heavy ropes as thick as a wrist are tied around the upper third of the pole. Care is taken not to damage the delicate hummingbird, or the heavy beak of the eagle that protrudes lower down; the dorsal fin of the wolf-headed blackfish that swims the length of the pole must also be treated with care (a human face peers from its blowhole). At the base of the pole, nestled securely between the eagle's wings, wearing a tall spruce root hat and holding a canoe paddle, is the Steersman himself.

The ropes radiate out from the horizontal pole like ribbons in a primordial maypole dance, and sturdy planks have been laid on angle to guide the butt into place. There are dozens of people holding each line, waiting for instructions, and it is now that a leader emerges. Standing on top of the pile of excavated sand and dirt is Skilay's son, a young man doing his best to rise to a daunting occasion. He gives the call to haul back and the crowd surges toward the beach; the lines go taut and the pole grinds stubbornly across the ground. This is how it would have been to haul a whale ashore by hand. With another mighty heave, the far end of the pole lifts slightly and the butt slides down into the hole, splintering the planks as it goes. The sound is almost sickening, and it brings home the gravity of the task at hand; the crowd is so large and dense that if the pole should fall or roll, someone, maybe a number of people, will certainly be crushed or killed. But now there is no turning back, and carefully, arduously, the pole is hauled upright. There comes at last a moment when the pole is centered in its hole, supported only by the people who surround it, that it becomes clear to some what it means to be Haida—and plain to all how many hands it takes to resurrect a tree.

CHAPTER THREE

Wildest of the Wild

The Haida at Kiusta saw it first as a white spot on the horizon that slowly grew larger. The people were afraid and donned their dancing costumes. They began to dance to drive it away, but it continued to approach. The spot became a giant web, and in the distance they saw spiders crawling up and down the webbing. As the web came closer they could see it was attached to a boat, but no ordinary boat, for it appeared to have wings which flapped up and down in unison against the water. The spiders, it turned out, resembled human beings, except they had white faces. The Kiusta people believed the Santla ga haade—the ghost land people—had returned from the dead.

— William Matthews, former chief of Old Masset, via Margaret Blackman

The fierce character of the natives would, however, render any attempts at permanent settlements, unless in strong parties, dangerous. In one sentence, to conclude, these islands are more interesting to the geographer than to the colonist; to the miner they may be valuable, but to the agriculturalist they are useless.

—from THE HAIDAH INDIANS OF QUEEN CHARLOTTE'S ISLANDS, a report by James Swan, 1873

Four years before Captain Cook arrived on the Northwest Coast, a Spanish explorer named Juan Pérez Hernández weighed anchor at Monterey, California, then the northern limit of Spanish settlement, and sailed north into *aqua incognita*. His mission: to claim the entire Northwest Coast for Spain. Poor weather and fog kept Pérez's 82-foot corvette *Santiago* well offshore for the entire trip, and after five weeks of wandering through the heaving miasm of the North Pacific, the crew was dangerously low on food and water and showing signs of scurvy. As a result, they were forced to turn back well short of their goal of the 60th parallel—then the southern limit of Russian settlement in North America. By all accounts the voyage was a dismal failure, save for one historic encounter. Posterity was the last thing on these sailors' minds when, on July 18, 1774, a lookout spotted land. What lay before them was not the mainland of the recently expanded New Spain, as their captain supposed, but a minor island in an uncharted archipelago. Without knowing it, Juan Pérez and his crew had discovered what would come to be known as the Queen Charlotte Islands.

While probing the coastline of what is now Langara Island, the *Santiago* was met offshore by a number of Haida canoes; as the paddlers sang, a shaman aboard the lead canoe scattered eagle down upon the water in front of the mysterious craft. Meanwhile, two priests aboard the *Santiago* admired their pleasant manner as well as their surprisingly fair skin and rosy cheeks; they also noted that one of the canoes held a spear tipped with iron. Where, they wondered, had these pagans acquired such a sophisticated item? It wasn't clear whether this weapon was for spearing otters or enemies, but the Haida seemed friendly, and some informal trading began, in the course of which they eagerly invited the sailors to come ashore. Sixty miles southeast of their position stood the golden spruce. By now the tree would have been about seventy-five years old and a hundred feet tall.

One can only wonder what the Spaniards, so obsessed with precious metals and so willing to see heavenly portent in every twitch of the landscape, would have made of a golden tree in a green forest. We will never know because the wind died and a strong current bore the ship away. This may have been for the best: up until this time, no explorer who set foot on the northwest coast had ever made it back to his ship.

This was only one of the reasons the Northwest Coast was such a late addition to the world map; with the exception of the two poles, this was the last significant feature to be added to the earth's portrait. There were two primary reasons. One was motivation: there wasn't any. Although places as tiny and remote as the Spice Islands were internationally famous by the sixteenth century, the islands of the North Pacific, and any riches they might contain, were unknown to Europeans. The other reason was access: there was simply no direct route; even Tasmania was easier to get to. An overland journey from the Atlantic was not only extraordinarily dangerous, but it could take years, and traveling from Europe presented an even more daunting prospect. During the 1720s it took the naval explorer Vitus Bering three years just to get from Moscow to the Pacific in order to begin his voyage; even then, the yet-to-be-named Bering Strait and most of Alaska still stood between him and the Northwest Coast. The sea offered no better alternative. Unless you were sailing from the east coast of Asia, the only way to get to the North Pacific involved detouring to the opposite end of the planet and passing around either South America or Africa (depending on one's direction of travel).

The Chinese, who had ships capable of crossing the Pacific by 1200, wrote of a legendary place called Fousang, which is believed to have been the Northwest Coast. The English had a less elegant name for it: they called it "the backside of America," and until it was accurately mapped at the end of the eighteenth century, it was subject to a series of cartographic indignities driven by misinformation, wishful thinking, and bald-faced lies. Quivira, the legendary city of gold sought by the Spanish during the mid-sixteenth century was rumored to be there, along with various lost cities, the Northwest Passage and its mythical precursor, the Strait of Anian. While writing *Gulliver's Travels*, the satirist Jonathan Swift chose the poorly understood region as a suitable location for Brobdingnag, the land of giants. *Gulliver's Travels* was published in 1726, two years before Bering tested the important, if rudimentary, theory that Asia and North America were actually separate continents.

The first European to set foot on the Northwest Coast and survive an encounter with the locals was Captain James Cook, who stepped ashore at Resolution Cove on the northwestern shore of Vancouver Island, on March 29, 1778. Vancouver Island is the biggest piece in the coastal puzzle; with its southern end nestled into a pocket formed by Washington's Olympic Peninsula, it angles northwestward off the B.C. coast for 300 miles. Cook's purpose for landing there would prove prophetic: he needed logs. While en route from New Zealand, via Hawai'i, both his ships sustained heavy damage to their masts and spars. The explorers' host on Vancouver Island was the powerful Nuu-chah-nulth chief Maquinna. He and his people wore cloaks made of sea otter skins and they lived in wooden houses that would have been recognizable to any European. Made of straight planks, with a smoke hole centered along a symmetrically peaked roofline, Nuu-chah-nulth bighouses would have appeared unusual only because of their great size and massive timbers. In addition to finding sea otters "as plentiful as blackberries" and gracious hospitality that augured well for future trade, another opportunity revealed itself deeper in the forest: trees the likes of which no Englishman had ever seen—an empire builder's fantasy. But Cook, bound again for Hawai'i, wouldn't live to see his discovery come to fruition.

When Cook's account of his third and final voyage was published in 1784, explorer-entrepreneurs, who had certainly heard rumors beforehand, took note and wasted no time in outfitting ships for the North Pacific. By 1785 the first vessel was on the coast, trading with the Indians, and nothing would be the same there again. These acolytes of Cook were called Nor'westmen (both the men and their vessels went by this name), and they were commercial explorers engaged in what were arguably the most ambitious, far-ranging, and culturally complex trading missions ever routinely undertaken. Their sole motivation was the pelts of a small sea mammal which had been classified not long before as *Enhydra lutris*, otherwise known as the sea otter. Their skins were the Golden Fleece of the North Pacific; the Chinese were paying a fortune for them. The Manchu dynasty of the eighteenth century, ruling over what they called the Celestial Empire, was the most advanced civilization on earth. With its enormous land area, the xenophobic society's 300 million citizens (then more than a third of the world's total population) were largely self-sufficient. An exception was sea otter pelts which members of the upper class coveted above all other clothing; as much as 120 Spanish silver dollars might be paid for a single high-quality skin, the equivalent of about $2,400 today. So precious were these furs that crewmen on trading vessels had their belongings searched periodically to make sure they weren't smuggling skins for their own gain, just as African diamond miners are searched today. While the East Coast cod, timber, and fur trades had been generating wealth for a century or more, the otter trade was the first northern commodity to send its exploiters into a bona fide frenzy, like gold, oil, or drugs.

With traders approaching by both land and sea, it was the fur trade that first opened the West; beaver, fox, and ermine were all profitable, but sea otter skins were in a class by themselves. Alexander Mackenzie, a fur trader and partner in the British-owned North West Company, was the first European to cross the continent overland, arriving on the coast in 1793, directly opposite the southern tip of the recently named Queen Charlotte's Isles (his journey was so arduous that no one has been able to duplicate it). Though he preceded Lewis and Clark by more than a decade, Mackenzie arrived to find dozens of ships already prowling the coast in search of otter skins, and many of them were American; as early as 1791, coins minted by the Massachusetts Bay Colony were seen dangling from the ears of North Coast Indians. John Jacob Astor, whose vast fur-trading empire has become the stuff of American legend, didn't send his first expedition until nearly twenty years later (1810). By then the slow-breeding sea otter was already in decline.

The sea otter, which exists only in the North Pacific, is unique among mammals; whereas the human head may be covered with 100,000 hairs—total, a sea otter can produce up to 600,000 hairs per square inch. So fine is their fur that it can be brushed in any direction; the result is a pelt of unparalleled softness. Lacking the insulating fat of other marine mammals, it is this dense mat of filaments, which the animals manually load with heat-retaining air bubbles, that allows these creatures to survive in North Pacific waters. Sea otters seldom go ashore, preferring to eat, sleep, relax, and copulate while floating on their backs. They carry flat stones in skin flaps under their forelegs which they place on their chests and use like anvils for breaking open shellfish (these stones are confiscated in aquariums because their owners will also bang them against the glass walls of their tanks). Sea otters are famously playful and affectionate, and they may float for hours holding "hands" with another otter. Mating, however, is a joyless exercise preceded by the male grasping the female's snout in his teeth and flipping her, belly up, onto his stomach. Apparently, they were extremely easy to kill.

The otter traders' route took them, literally, around the globe on a highly profitable journey that came to be known as the Golden Round. While some traders embarked from colonial bases at Macau and Calcutta, many others shipped out from their home ports in the North Atlantic; from there, it took three or four months just to reach Cape Horn, a gauntlet of fog, icebergs, gale-force winds, and huge waves, all of which flow in the opposite direction of Pacific-bound ships. Square-riggers weren't designed to sail into the wind, and for this reason it might take a month of tacking just to get around the Horn, an ordeal that exacted a heavy toll from ship and crew alike. Some captains simply gave up and turned their vessels around though one fur-mad trader made the journey in a 33-foot schooner. From

Cape Horn, just shy of the Antarctic circle, these vessels would tack northward again for 8,000 miles until they reached the thick fog, fickle winds, and ferocious currents of the Northwest coast. It was here, after half a year of hard sailing in cramped, vermin-infested conditions, that the real work began, and there was no respite for travel-weary sailors. The coast's overwhelming dampness not only led to frequent respiratory ailments but it rotted food, canvas, and rope at alarming speed. Poor visibility, along with fluky winds, whirlpools, and bizarre tidal upwellings made for hazardous coastal navigating. In some surge channels, tides run at speeds only slightly slower than Niagara Falls. Frequent losses of anchor and chain due to the region's rough "holding ground" led one captain to recommend embarking for the coast with no fewer than five spare anchors and cables. The chronically foul weather further demoralized the crews, who employed a veritable thesaurus of gloomy descriptors to express their experiences there: "dreary," "inhospitable," "wretched," "savage," "barbarous," and the "wildest of the wild" being but a few. Some of their shipboard experiences sound like they were conjured up by Hieronymus Bosch: Ice-cube-sized hail would cause birds to drop, dead, from the sky. One sailor compared the seasickness he and his mates suffered en route to "shitting through one's teeth."

Once a load of otter skins had been procured, a ship would head south to Hawai'i for resupply, ready sex, and perhaps an ancillary load of sandalwood. From there the ships would cross the Pacific to Canton, braving both Asian and European pirates along the way. (The Russians, who had a half-century head start on the Europeans, sent their furs overland, mainly through the town of Kiakhta on the northern Chinese border.) All profits from the sale of skins would be reinvested into Chinese tea, silk, and porcelain. Once reloaded, the Nor'westmen would make their way south into the Indian Ocean, around the Cape of Good Hope, and back up through the Atlantic to their home ports. A typical "round" might take two years and cover more than 40,000 miles, during which the Nor'westmen emptied and reloaded their holds twice and traded with dangerous and widely divergent peoples speaking a minimum of four unrelated languages. In addition to English and French, the Chinook trading jargon was popular on the south coast as far as Vancouver Island while a Haida-based equivalent came in handy up north. Someone on the ship also had to be familiar with Hawaiian and Cantonese.

Meanwhile, up and down the coast the locals were having surreal first contact experiences with strange craft inhabited by beings who could do things that no ordinary human should be able to do: they could remove the tops of their heads at will (wigs); they could shed their colorful skins and pull objects out of their bodies (close-fitting clothes); they had weapons that could pierce battle armor made from wooden slats and the thick hides of sea lions. And their eyes were blue. As the Indians got to know them better, these aliens became known, first, as Iron Men and, more specifically, as Boston and King George Men. They appeared to be of only one gender; with the exception of the occasional captain's wife or Hawaiian mistress, there were rarely women aboard. However, this and their peculiar smell were overlooked because they carried with them a wide variety of amazing items they seemed eager to part with, including chisels, nails, copper pots, scissors, mirrors, buttons, blankets, and brass bells. But traveling with them, too, were the Four Horsemen in the form of rum, guns, contagious diseases, and a strident worldview. These visitors from afar were not, it turned out, returning from the land of the dead; rather, they were bringing it with them. Within a century a stranger traveling the West Coast and seeing, firsthand, village after village strewn with the bones of the unburied might have reasonably supposed that the Land of the Dead was right here, in North America. It wasn't that the people of the Northwest Coast were strangers to murder and mayhem, or even to disease—not by a long shot: the Haida took their enemies' heads, after all, and smallpox almost certainly preceded the traders. It would be the scale, as well as the range, of devastations that they would find so overwhelming.

Without a doubt, the advantages of novelty and surprise did give the foreigners the upper hand in the first rounds of trade. Some of

Cook's men, for example, realized a profit of 1,800 percent on the otter skins they procured from the Nuu-chah-nulth, thereby setting off a near mutiny among some of his crew who wanted to abandon their "voyage of discovery" and head back to the coast for more skins. However, the natives quickly reassessed the value of these new trade items in relation to their own, and from then on every deal became a game of wits.

As much as they might have liked to believe otherwise, the new world the Nor'westmen found themselves in was not an innocent or naive place by any stretch of the imagination. Intertribal trade was well developed by the time the foreigners showed up, and a wide variety of goods, ranging from copper and puffin beaks to human slaves and the scalps of woodpeckers, were finding their way back and forth from California to Alaska, and from the outer islands to the Great Plains. The newcomers learned, much to their annoyance, that such fundamental laws and practices as supply and demand, false advertising, gouging, circumventing the middleman, not to mention the old bait and switch, were already in wide usage on the coast. As one Nor'westman put it, "These artists of the northwest could dye a horse with any jockey in the civilized world, or 'freshen up' a faded sole with the most ingenious and unscrupulous of fishmongers." Throwing the horny all-male crews still further off-balance was the fact that North Coast women, who tended to be less sexually available than the Hawaiians, often played a major role in trading negotiations.

For well over a hundred years, there has been a strong tendency throughout much of the Northern Hemisphere to idealize the American Indian; this extends, in many cases, to the Indians themselves. They are often depicted as proto-environmentalists—stewards of a continental Eden who revered their prey and nurtured the land until it was laid waste by invading Europeans. Such rose-tinted hindsight is surprising given that so much information about the realities of tribal life survives to this day. But it wasn't just the likes of John Muir, Edward S. Curtis, and James Fenimore Cooper who subscribed to this view; even George Armstrong Custer, of all people, was known to wax rhapsodic about the passing of what he and many of his contemporaries called the "noble race." And yet, before the westward expansion, before any of these romantics was yet born, the West Coast otter trade was helping to set the tone for every extractive industry that has come after.

Though food was generally plentiful on the Northwest Coast, natives would certainly have been familiar with hunger and hard times in the form of bad winters and poor fish runs. While it was not a food item, the sea otter provided some of the finest clothing available anywhere. And yet, despite its practical importance, and despite a necessarily keen sensitivity to the rhythms of the natural world, the West Coast Indians pursued this creature to the brink of extinction. In doing so, they demonstrated the same kind of profit-driven short-sightedness that has wiped out dozens of other species, including the Atlantic salmon and, more recently, the Atlantic cod. It is an eccentric and uniquely human approach to resources: like plowing under your farmland to make way for more lawns, or compromising your air quality in exchange for an enormous car.

From the vantage point of the twenty-first century, it is hard to say who was more inebriated by greed: the Europeans who were seeing profits in the hundreds of percent, or the natives who were suddenly able to leapfrog their way to the top of the social hierarchy and put on spectacles of largesse hitherto unimaginable by any potlatch host on the coast. So eager were the Indians to get their hands on the traders' various technological marvels that a man would readily sell the otter cloak off his wife's back and, on occasion, her back as well. And so desperate were the Iron Men to acquire these skins that they would trade away virtually anything that wasn't crucial to the journey home; this included Indian slaves from down the coast, firearms, silverware, door keys, and the sailors' own clothing. These were boom times for all concerned, a rapacious festival of unrestrained capitalism.

Despite the claims of western movies and popular history, the West, in fact, went "wild" seventy-five years before the arrival of the railroad, Jesse James, or the Seventh Cavalry. By the time Lewis and

Clark arrived on the Pacific coast in 1805, the local Indians were already heavily armed. As early as 1795 the Haida were returning the traders' cannon fire with cannonades of their own—the guns having been pillaged from captured European ships. By 1810 some chiefs possessed such formidable arsenals that they were selling top-of-the-line swivel-mounted cannon *back* to the Nor'westmen. The Haida reportedly had such weapons mounted on the bows of their canoes. The Indians had understood the art of fortification since well before the arrival of Europeans, and at least one Haida village near Masset had a stockade armed with plundered cannon. Further north, the Tlingits were taking measures of their own; annoyed that the Russian American and Hudson's Bay Companies were encroaching on their role as middlemen between inland and other coastal tribes, they reduced the Russian and British forts to smoking ruins. Meanwhile, of the dozen or more trading vessels taken by West Coast Indians before the collapse of the otter trade in the 1850s, fully half were seized by the Haida.

AN EARLY SOURCE OF TENSION emerged around the fact that theft was an accepted practice among virtually all the people the traders encountered. To say that natives would take anything that wasn't nailed down was, apparently, an understatement: John Meares, one of the first of the Nor'westmen, reported that "it has often been observed when the head of a nail either in the ship or boats stood a little without the wood, that they [natives] would apply their teeth in order to pull it out." Thefts were perpetrated with a sporting attitude, similar to the Plains Indian practice of counting coup; the assumption seemed to be that whatever you couldn't effectively protect—whether it be a soupspoon or a schooner—you didn't deserve to own in the first place. The white traders were, of course, practicing their own version of this: while the natives were making off with tools, laundry, and rowboats, traders thought nothing of coming ashore and helping themselves to water, timber, and game—all items that natives considered to be their property.

Because of this, deals were often brokered in an atmosphere of mutual suspicion and contempt barely hidden beneath a thin veneer of carefully orchestrated protocol: gift giving, invitations to dine and visit the other's living quarters, etc. However, as competition and inflation grew with explosive speed, it didn't take long for the presents and feasting to degenerate into tense, heavily armed encounters that bore a strong resemblance to a contemporary drug deal or hostage exchange. Much of the character of an individual transaction came down to the personalities of those involved, and there were honorable square dealers to be found on both sides. But bad news travels fast, and one man responsible for the early and rapid deterioration in trade relations was James Kendrick, who will go down as one of the most destructive (and prophetic) trade ambassadors in early American history. Kendrick was, among other things, the first man to sell large quantities of arms to the West Coast Indians, including the Haida, and it is thanks in part to him that the Queen Charlotte Islands have the bloodiest history of any place on the coast.

Things might have turned out differently if Captain Kendrick, one of the "Boston Men," hadn't had his underwear stolen one day in June 1789 and decided to teach the local chief, Koyah, a lesson by sticking his leg in a cannon barrel, cutting off his hair, and painting his face. This was a devastating humiliation for Koyah, a renowned and wealthy chief, and restoring his lost status became an obsession. When Kendrick returned two years later, Koyah was waiting for him; he managed to capture Kendrick and his ship, but was outgunned in the end. A massacre ensued in which as many as forty Haida were killed and scores more were wounded (the battle was later commemorated in a broadside called *The Ballad of the Bold Northwestmen*.) Koyah survived, and the next ship to visit his territory was burned to the waterline and her crew slaughtered, with the exception of one man who was enslaved. That same year, one of Koyah's allies gave another vessel the same treatment. In 1795 Koyah led an attack of more than forty canoes carrying approximately 1,200 warriors against yet another American ship called the *Union*; the assault was repelled by overwhelming force and as many as seventy Haida were killed. "I could have kill'd 100 more with grape shot," wrote the *Union's* twenty-year-old captain, "but I let humanity prevail & ceas'd firing . . . None of us was hurt."

Chief Maquinna, the same man who had given Captain Cook such a warm welcome, was driven to a similar extreme. Five years after Cook's visit, he was paid a call by the *Sea Otter*, the first fur-trading vessel on the coast. Maquinna was invited aboard and shown to a seat of honor that had been booby-trapped with a charge of gunpowder. The chief was subsequently blown out of his chair; he survived but was scarred for life. When Maquinna's warriors attacked in retaliation, dozens were killed by gun and cannon fire. On other occasions, traders ransacked Maquinna's home and summarily executed his subchiefs. Nearly twenty years after the exploding chair incident, Maquinna oversaw the seizure of the *Boston* and the massacre of all but two of her crew; only the highly valuable armorer and a fortunate sailmaker were spared.

Of all the sticky endings met by Nor'westmen, Captain Kendrick's may have been the most poetically just. In 1795, six years after his battle with Koyah, the moody, alcoholic Kendrick was in Honolulu Harbor where he requested a cannon salute from a British ship called the *Jackal*. The *Jackal* obliged—accidentally, with live ammunition—and James Kendrick went down in a hail of grapeshot. A month later the *Jackal's* master was killed by Hawaiians. Kendrick's brother was subsequently killed by an ally of Koyah's.

WHILE THE LOCALS could always retreat to their village forts or, if worse came to worst, into the deep forest, the Nor'westmen had nowhere to go but their ships, and at anchor they were sitting ducks. Traders reported being surrounded at times, by *hundreds* of canoes, some of which would have been longer than the ships themselves;

they were also far more maneuverable in close quarters. Escape, under these conditions, would have been impossible, and there was always the possibility of attack, even in the most apparently benign situations. William Sturgis, a veteran fur trader from Massachusetts who would become one of the harshest critics of his colleagues' behavior on the coast, created a formula for an efficient, nonviolent trading environment. His recipe for success was, in short, a seamless defense coupled with a compelling display of ready firepower.

It is hard to overemphasize the importance of these vessels to the sailors whose lives depended upon them. The journeys these men were engaged in would be considered epic today; for all practical purposes, they were closer to being interplanetary than intercontinental. Like a spaceship, each vessel was a life-support system unto itself, serving as dormitory, mess hall, clinic, storefront, warehouse, boardroom, fortress, armory, and escape module rolled into one. Without it, there was no way home. If something went wrong en route, you would probably die; there was no way of calling for help, and rarely anyone to hear you even if you could. In the event that your ship was lost and you managed to make it to shore, it would, in most cases, merely prolong your suffering. A sailor separated from his mother ship was an extremely vulnerable individual; he stood an excellent chance of being killed outright or enslaved by people who were alien in every sense of the word. The difference between his experience and that of contemporary West African slaves would have been only a matter of scale.

In retrospect, it is difficult to fathom why the traders were so willing to arm the Indians, particularly when one considers that, as one French trader observed, the Indians would frequently turn their weapons on the very men who sold them, and on the same day they had been acquired. (The Spanish had a policy of never trading arms with natives.) In some cases, the sale of guns was intended to buy loyalty, as was the case with British fur traders who had cut arms deals with several Plains Indian tribes. Because they often traded inferior weapons, some traders may have felt confident that they could always outgun the Indians if it came to a fight; others probably thought they would never return to the area so it didn't matter what they left behind. Or maybe they just weren't thinking. In any case, the speed with which the Indians adapted to new technologies and changing conditions caught many of the traders by surprise.

IT SEEMS IMPOSSIBLE that participants on either side of the fur trade would have failed to envision for the sea otter the same fate that John J. Audubon could see awaiting the buffalo in 1843, when vast herds still blackened the plains. "Before many years," wrote Audubon in his *Missouri River Journal*, "the Buffalo, like the Great Auk, will have disappeared; surely this should not be permitted." In 1730 millions of sea otters flourished in the kelp beds that dotted the Pacific coast, from Baja California north to Alaska, and south again along the Aleutian Islands and Kamchatka, all the way to Japan; by 1830 the species had been all but extirpated from most of its range. And yet, while the Indians appear, in most cases, to have been willing, even zealous, agents of the species' destruction, the white traders had them over a barrel. Coercive techniques, including threats and hostage taking, were used by some traders, but in a sense the natives were hostages, first and foremost, to the trade itself: once the market for skins had been created, they really had no choice but to participate. Any village or tribe that didn't would become the losers in the inevitable race for new arms, technology, and wealth. Once aboard a juggernaut like this, it appears suicidal to jump off—even if staying on is sure to destroy you in the end.

As the sea otter population dwindled, intertribal warfare grew so vicious, and trade relations soured so completely—on all sides—that commercial ventures were no longer worth the risk. Increasingly mutinous crews as well as the kidnapping and ransoming of natives for skins further exacerbated the situation. William Sturgis, who lost a brother to the Haida, gave what may stand as the fairest assessment of the situation on the coast at the start of the nineteenth century. Recalling his experiences in the otter trade, he wrote:

> Should I recount all the lawless & brutal acts of white men upon the Coast you should think that those who visited it had lost the usual attributes of humanity, and such indeed seem to be the fact. The first expeditions were . . . entrusted to such men as could be picked up ready to undertake a hazardous adventure. These were often men of desperate fortunes, lawless & reckless, who, upon finding themselves beyond the pale of civilization and accountable to no one, pursued their object without scruple as to the means, and indulged every brutal propensity without the slightest restraint. . . . I do not exaggerate when I say that some among them would have shot an Indian for his garment of Sea Otter skins with as little compunction as he would have killed the animal from whom the skins were originally taken.

The quick and dismal failure of trading relations on the Northwest Coast can be traced to a pair of lethal ingredients: the fact that both parties brought extremely violent cultures to the bargaining table, and that neither side was willing to see the other as fully, "legitimately" human. This combination of violence and disdain, coupled with a strong sense of entitlement, helped set the tone for future settlers' and investors' attitudes, not just toward the New World's human inhabitants, but toward its resources as well. Little, in fact, has changed since King William III declared from an ocean away that the forests of Maine were "the King's Pine."

While the sea otter "gold rush" was capturing people's imaginations and poisoning them with greed, cooler heads were noticing a commodity that would prove far more lucrative over the long term. In 1787 Captain John Meares, who could be considered the father of the Northwest timber trade, received the following orders from his backers in London: "Spars of every denomination are in constant demand here. Bring as many as you can conveniently stow." A year later, his decks stacked with Vancouver Island timber, Meares himself was moved to write, "Indeed the woods of this part of America are capable of supplying . . . all the navies of Europe." It may have been the sea otter that brought them, but the timber is why they stayed.

CHAPTER FOUR

The Tooth of the Human Race

I am the tooth of the human race,
Biting its way through the forest vast,
Chip by chip, and tree by tree,
'Til the fields gleam forth at last,
Eating its heart with keen delight,
Into the groaning tree I bite.
Every stroke the land doth bless,
And joy o'erflows the wilderness.

—Donald A. Fraser, "The Song of the Axe,"
adapted by Margaret Horsfield

JOHN MEARES MAY HAVE had a vision, but he wasn't the first European to look at the New World and see a navy in the trees. Everyone who set eyes on the North American coast, from Columbus and Cabot forward, noted the vast quantities of timber, but the English were the first to systematically exploit it. Like the

Romans, Greeks, and Sumerians before them, the English had an insatiable appetite for wood; as a result, the thickly forested British Isles had been reduced, largely, to pastureland before Meares was born. By the time he made captain, the British Empire was the closest thing to a superpower the world had ever known, an achievement due, in large part, to her formidable and far-ranging navy. Wooden ships pioneered global trade and transoceanic empire building, but they did so in part to perpetuate themselves (a rough rule of thumb for gauging the timber needed to build a late eighteenth-century warship was an acre of oak forest per cannon). Tall, knot-free pine for masts and spars had become hard to find in western Europe, and it was for these that royal shipwrights turned to North America. Up until a hundred and fifty years ago, a forest of straight, sturdy pine was as valuable as an oil field or a uranium mine today: it was a critical source of energy (i.e., sail power) without which a nation could not fully realize its commercial or military ambitions.

By the time Captains Cook and Meares arrived in the North Pacific, agents of the British Crown had already been logging the "pineries" of eastern North America for more than a century. Ships' masts were one of the New World's first significant exports, along with cod, potash (derived from charcoal), and beaver pelts. As early as 1605, samples of white pine from Maine were being sent back to England for testing by the Royal Navy, and by 1691 England's "Broad Arrow Policy" was in effect. Reflecting the wholesale audacity of the times, this highly unpopular decree stated that any trees twenty-four inches or more in diameter located within three miles of water were automatically the property of the king. Lest there be any confusion about whose woods these were, the royal mark of the broad arrow was blazed into their bark. The marked trees were considered so valuable that mast ships—custom built to accommodate long timbers—traveled in convoy with armed escorts.

Three hundred years on, such zealous precautions seem almost quaint, and yet they offer a graphic measure of the true value of wood, a substance whose importance in our history and evolution is almost impossible to overestimate. Throughout most of the world, and for most of human history, wood has been the principal source of fuel and building material, providing heat, light, and shelter as well as food, clothing, and weapons. Nowhere is this dependence more vividly evident than in North America. Trees, it could be said, represent the bones of our collective body. So central have they been to our existence that an archaeologist examining the iconography of New England settlers in the seventeenth century might reasonably suppose that these devout Christians were really druids or had simply "gone native." In 1652, after decades of using a motley assortment of currencies, ranging from Indian wampum to tobacco and Spanish silver, the Massachusetts Bay Colony began minting its own money. These crude coins were not decorated with crosses, kings, or familiar symbols of liberty, they were embossed with trees, specifically, pine, oak, and willow. "What better way to portray the wealth of our country?" wrote the coins' die maker, Joseph Jenks. Likewise, early "American" flags were not the star-spangled banners we are used to seeing, but, rather, banners honoring the tree. New England's first flag looked much like Vermont's does today; so did the flag borne by George Washington's cruisers; the Bunker Hill and Continental flags followed the same theme. There was even a Liberty Tree flag. In some cases, the flags themselves were made of wood. From upstate New York to Florida and Texas, the United States is still dotted with "treaty oaks" where historic agreements were signed between early settlers and local Indians. Trees were the continent's first churches, capitol buildings, and fortresses, and their iconic importance—like that of the golden spruce for the Haida—has persisted into the space age: the Canadian maple leaf flag dates only to 1965.

But the reverence that trees engendered was not always extended to the forests they came from; most New World settlers arrived from pastoral places long since cleared for agriculture and grazing land, and a boundless treescape thriving with unfamiliar people and animals was a shock. It wasn't just the continent's scale they found so overwhelming, but its dense and endless secrecy: the forest is an introverted wilderness, and it offers risk and refuge in equal measure. Robin Hood found sanctuary there, but so did Red Riding Hood's Wolf (who, in the end, was killed by a woodcutter). While the armies of empires dominate the open plain, rebels and patriots gain advantage in the shelter of the trees—right beside outcasts, outlaws, and mystics. The woods provide food and building materials, and yet they also disorient and impede progress. Until relatively recently, North American staplefood species like deer, elk, bison, and caribou inhabited the forest from coast to coast, but so did wolves, bears, and mountain lions, creatures that continue to fascinate, terrify—and kill us—to this day.

Before the arrival of Europeans, the Indians used fire as an effective, if haphazard, method of driving game and opening up these great forests for cropland and "pasturage" for game animals, but those sundappled parklands extended only so far. Although most North American tribes made their homes in or near the forest, virtually all of them told stories of a foul-smelling, flesh-eating monster that lurked in the woods beyond the village. Stories like "Hansel and Gretel" are the Old World equivalent; the 1998 film *The Blair Witch Project* succeeded in part because it tapped into these same deeply rooted fears. In his book *Of Plimoth Plantation*, published in 1651, the Pilgrim William Bradford described the low forests of Cape Cod as a "hideous and desolate wilderness, full of wild beasts and wild men." He was not alone; for many of the early settlers, clearing the land was not just a necessity, it was a sacrament—an act of holy alchemy in which the dark, evil, and worthless was transformed into something light, virtuous, and fruitful. Profitable, too. Once those settlers who hadn't fled back to England moved out of their (literal) holes in the ground, learned to use the canoe and grow local crops (usually in Indian fields), the entrepreneurial spirit wasted no time in elbowing the Holy Spirit aside. Fortunes were made on wood exports to denuded England and Spain as well as the West Indies. By 1675 hundreds of sawmills were operating throughout New England and Atlantic Canada.

LOGGING IS AN INDUSTRY THAT, while unseen by most of us, has altered this continent—indeed, *all* inhabited continents—even more completely than agriculture. This has been the case, not since 1865, or 1620, or 1066, but for millennia. Logging is the prerequisite to life as we know it. first and foremost, the trees must go. In this sense, the woodcutter has been the pointman for Western civilization (indeed, all civilizations). Not only has he imposed a tidy, "rational" order on Nature's apparent chaos, but he has provided the space and materials that have allowed us to feed and build our society, and to spread its message to the farthest corners of the globe. In fact, it has often been the quest for still more wood that has led us there.

If one were to encapsulate the entire history of Western logging into a thirty-second film, its effect on the Northern Hemisphere would be comparable to the effect of the eruption of Mount St. Helens on the surrounding forest: both represent irresistible waves of energy that originated in a relatively small, specific area and expanded rapidly, leveling everything in their path. The earliest known references to western logging date from around 3000 B.C.; they came from the pro-tourban centers of Babylonia (what is now Iraq). There, in the so-called cradle of civilization, logging of a kind that we would recognize today (i.e., the cutting and trading of wood for commercial and nation-building purposes) built cities and navies before spreading steadily westward, ever faster, across Asia Minor and Europe until it reached the Americas, where the pace would quicken to a sweeping blur. Left behind are landscapes we take for granted, though they bear scant resemblance to their preagricultural states. The Lebanese flag has a cedar tree on it because much of what is now desert was thickly forested before the harbingers of civilization—i.e., woodcutters, farmers, and goats—saw to it that large stands of cedar will never grace the Holy Land again. The stark and sere limestone hills that we think of as typical Greek and Italian landscape were once all but invisible beneath a layer of long-gone topsoil held in place by forests of cedar and oak. The pastoral idyll that is rural Europe was once a pillared and leaf-domed shadowland inhabited by bears, wolves, and tribespeople who held the forest to be sacred. Those witch- and fairy-infested treescapes evoked so vividly by Shakespeare and the brothers Grimm actually existed, but, with the exception of a few forgotten pockets and a handful of parks, they have not been seen "in person" for hundreds of years.

Had they been available even two hundred years ago, aerial photographs of North America would have revealed a strangely familiar landscape. Instead of the contemporary tic-tac-toe board of browns and grays and greens whose uniformity is interrupted only by the occasional wrinkle of foothills and mountain ranges, North America would have reminded us of Dark Age Europe—or perhaps the Amazon. With the exception of the Great Plains and the desert southwest, the continent would have presented a virtually unbroken carpet of forest that stretched from the Atlantic Ocean to the Pacific, and from the Gulf of Mexico to the Gulf of Alaska. The total represents a nearly incalculable amount of wood—in the trillions of board feet—and yet the speed with which it has been cut, burned, and, in many cases, simply squandered—is unparalleled.

IT WAS ONLY A FEW GENERATIONS after Europeans began altering this landscape in earnest that an alternative view of the forest and its inhabitants began to emerge: salvation, claimed the Romantic philosophers and writers, lay not in a tame and planted landscape, but in the raw wilderness. But the proponents of these views tended to come from settled areas and knew little of the deep forest, or of the labor required to clear it. In 1864, when much of the New England wilderness had been brought to heel, one of Nature's most fervent eulogists, Henry David Thoreau, would still find the great Maine woods to be a little more "natural" than he had bargained for. Far from the comforts of suburban Concord, he came away shaken; this great, unkempt northern forest was "savage and dreary," he wrote, "more lone than you can imagine" and "even more grim and wild than you had anticipated."

Thoreau made these observations as North America's wood consumption was exploding. Between land clearing, fuel burning, and construction, spectacular quantities were being devoured by a continent in the throes of full-blown expansion. The industrial revolution, combined with rapidly expanding settlement and a deluge of immigrants, accelerated the process of digesting forests exponentially. The circular saw—the whirling heart of every North American sawmill—had been introduced from England in 1814. 1828 saw the arrival of the planing machine, which allowed for the rapid manufacture of floorboards. Five years later, balloon construction (the fast, cheap, and simple technique of building with two-by-fours covered with sheathing) was introduced in Chicago; it remains the most popular method for constructing houses today. Prefabricated housing followed shortly after, being used to shelter California gold rushers in the 1850s; by this time there were factories capable of producing an unheard-of one hundred panel doors in a single day, and multibladed "muley" saws that could reduce a whole log to a stack of boards in one pass. By 1840 there were more than thirty thousand sawmills, shingle factories, and related wood-processing establishments operating east of the Mississippi River (more than six thousand in New York State alone). Between 1850 and 1860, more than 60,000 square miles of North American forest was liquidated. In 1867, when timber was (briefly) America's second biggest industry—after cotton—one of the first inventions specifically designed for mass disposability arrived in the form of the paper bag. By 1900 North Americans were felling and clearing in excess of 50 billion board feet of timber per year.

The European settlers of North America mastered their environment as no one had before; not only were they logging the continent faster than anyone else in history, they were putting its wood to a more magnificent array of uses. So sophisticated were craftsmen in its many applications that by 1825 even something as simple as a chair might contain fifteen species of New World wood. Each kind served a specific structural or aesthetic purpose, and together they created a nearly seamless and synergistic whole with a versatility, durability, and strength-to-weight-to-cost ratio unequaled by any other building material. To this day, there is none to match it. Standing on the shoulders of imported technology, which was arriving almost daily in the minds and luggage of immigrants, New World inventors and craftsmen were transforming trees into everything from shoes, clocks, and sewer pipes to canyon-spanning trestle bridges and—eventually—airplanes and celluloid film.

While timber shortages had forced the British to rely increasingly heavily on coal throughout the seventeenth century, wood remained the dominant fuel in North America for another two hundred years. By 1870 eight million cords of wood were disappearing into the fireboxes of American locomotives annually—enough to build nearly 700,000 homes. Meanwhile, the iron furnaces of western Massachusetts were consuming 16 square miles of forest per year. In the same period, the sawmills of central Maine would generate a quarter of a million cubic meters of *waste* wood. It is estimated that a quarter of all the timber that passed through mid-nineteenth-century mills came out as sawdust, and all of it had to be burned for reasons of safety. Mills were usually located along waterways, and not only did the great quantities of sawdust and wood scrap create hazards to navigation, but these, along with logjams, would sometimes catch fire, causing rivers to burn for weeks, just as they would from oil and chemical pollution a century later.

Seasonal burning of cropland and forest scrub has been standard procedure since prehistoric times, and the incineration of sawmill scrap and forest slash only added to the acrid cloud hanging over much of the New World. So thick and persistent was this pall of smoke that it often paralyzed shipping traffic on major rivers. In 1868 lighthouses were recommended as a navigational aid on Oregon's Willamette River—not for the winter fogs, but for the autumn fires. Throughout the United States and Canada, logging practices transformed the forest itself into a major fire hazard. Where the buffalo skinners left small mountains of skulls and bones in their wake, the loggers left slash piles—jagged heaps of highly flammable forest trash that could be acres across and a dozen feet deep. Far more concentrated than naturally occurring forest fuel, slash piles were conflagrations waiting to happen, and when, inevitably, they did, the results were apocalyptic. Survivors frequently recalled their conviction that Judgment Day was at hand, and it was in reference to the most lethal of these holocausts that the term "firestorm" was coined. On the same day as the Great Chicago Fire, in 1871, the Peshtigo, Wisconsin, fire burned 1,200,000 acres (nearly 2,000 square miles) in twenty-four hours and killed an estimated 1,500 people—so many that hundreds of the dead were buried in mass graves because there was no one left to identify them. In 1886 the young city of Vancouver, then consisting of about a thousand wooden buildings, was burned to the ground by a runaway slash fire in what was conservatively estimated to be about forty-five minutes. In 1894 twelve towns were destroyed and 418 people immolated or asphyxiated in the Hinckley, Minnesota, fire. Survivors described exploding "fire balloons" and corkscrewing flames spinning with such force in their self-generated winds that they tore trees from the ground and sent them whirling through the blackened sky like blazing pinwheels. Another slash fire in Seney, Michigan, burned with such ferocity that it cauterized the ground. The Midwest is pocked with such "stump prairies," many of which remain virtual wastelands to this day.

Even at this late date, with much of the East and Midwest "slicked off," the forest was still perceived as "an enemy to be overcome by any means, fair or foul." The push to open the West, coupled with the sweeping cultural changes effected by the industrial revolution and post–Civil War urbanization led to the woods being viewed—and treated—with a kind of aggressive contempt; the noun "lumber" was itself derogatory, meaning anything useless or cumbersome. North American immigrants were a restive people who tended to view land less as a "place" than as a cheap commodity. They cut the forest the way they breathed the air—as if it were free and infinite.

This is easy enough to say from the vantage point of twenty-first-century North America, where the experience of clearing wild land by hand is virtually unknown, but the act of removing branches, trunks, and roots from even an acre of thick forest was backbreaking—sometimes heartbreaking—work. Estimates vary as widely as the terrain, but, roughly speaking, it would take two men a year to make a dozen acres of eastern forest "fit for the plough." Most of those trees were felled with an ax. This crude but effective tool originated in the Stone Age and yet it has remained in wide usage throughout the world ever since. In 1850 it would have been as ubiquitous as the telephone is today; just about everyone would have known how to handle one. Axes—along with chain saws—are still standard equipment for professional woodsmen, and they were used actively for tree falling—even on the West Coast—into the 1950s. But the "ax age," as one historian calls it, reached its zenith in the late nineteenth century, and the North American version represented its highest state of evolution. During one demonstration, a man named Peter McLaren hacked his way through a thirteen-inch gum tree log in forty-seven seconds. Dozens of manufacturers, competing with hundreds of styles, had elevated this humble implement from a mere tool to a potent—even sexualized—facilitator of manifest destiny. Model names were often an ax's only distinguishing characteristic, and many sound as if they were dreamed up by the same ad agencies that promote motorcycles and firearms today. Climax, Demon, Endurance, Cock of the Woods, Red Warrior, Hiawatha, Hottentot, Black Prince, Black Chief, Battle Axe, Invincible, XXX Chopper, Woodslasher, Razor Blade, Stiletto, Forest King, and Young American were just a few of the choices. One model, for sale in Vancouver, was called the Gorilla.

BY THE MIDDLE OF the nineteenth century, the boundaries between British and American territories (and forests) were painfully clear on the Northeast Coast, but they remained far more tenuous in the Pacific Northwest. After the Spanish had been factored out of the northwestern equation in 1795, Britain and the United States were left to divvy up this huge, unwieldy slice of the continental pie. Unable to agree on a boundary dividing the western portion of British Canada from the rapidly expanding United States, the two rivals settled on a kind of territorial joint custody. From 1818 to 1845, Oregon Territory, a vast area extending from present-day Oregon's southern border all the way to southeast Alaska, was declared an "area of joint occupancy." Thus, for almost thirty years the Queen Charlotte Islands were considered part of Oregon, despite the fact that they were a thousand miles from the Columbia River and a day's sail from the nearest land. Meanwhile, the British-owned Hudson's Bay Company was operating one of the West Coast's first lumber mills on the Columbia, 300 miles south of the current U.S.-Canadian border. The situation became intolerable and, in 1846, under pressure from President James Polk and his saber-rattling campaign slogan, "Fifty-four forty or fight!" the current boundary was determined by the Treaty of Oregon.* Eight years later the colony of British Columbia was created, but in 1858 the province was invaded by tens of thousands of American gold miners, who posed yet another threat to British sovereignty.

By this time, the West Coast otter trade was finished. The Nor'west-men didn't linger in those barren waters but quickly redirected their efforts toward seals and inland fur species. At the same time, masts and spars, harvested from the mainland coast during layovers, became an increasingly important part of the West Coast traders' cargo; much of it was sold at Hawai'i, which had by then become a major crossroads for Pacific whalers and traders. Meanwhile, after their heady, destabilizing ride on the fur-driven economic bubble, the Haida came down with a crash. The otter, it turned out, was more than a spirit relation and a source of clothing, it was a bellwether for the tribe. Once it was gone, the Haida were reduced to selling carv-

---

* 54°40' was the latitude at which Captain Pérez and his men grew too sick to carry on and turned around. It still stands today as the southern boundary of Alaska.

ings to passing sailors and trading potatoes with former enemies. While their steel weapons rusted and their European clothes turned to rags, a biological holocaust of smallpox, influenza, tuberculosis, and venereal disease swept up the coast and over the islands. The Haida and their mainland neighbors died by the tens of thousands; villages turned to ghost towns; the culture was changed forever. In less than three generations, a legendary nation of untold age had traded its first otter skins to Europeans, glowed with a feverish inten-sity it had never known before, and flamed out. Miners, missionaries, Indian agents, and settlers followed, but the archipelago wouldn't attract the world's attention again for nearly a century. Next time, they would come for the trees.

For now, there was more than enough wood down south to keep the newcomers busy. In fact, it was almost too much of a good thing. Both the coastal forests and the country in which they grew were so grossly out of scale in comparison to anything the pioneers had seen previously that many were at a loss as to how to proceed. "The great size of the Timber and the thick growth of the underwood have been sadly against us in clearing the ground," wrote James McMillan, founder of Fort Langley, which was built in 1827, thirty miles upriver from present-day Vancouver. "[T]he jungle on the banks of the [Fraser] River is almost impenetrable and the trees within are many of them three fathoms [eighteen feet] in circumference, and upward of two hundred feet high."

"When I stood among those big trees," wrote an American pio-neer woman shortly after her arrival on the coast, "I felt so afraid, of what I do not know. Just afraid."

"I raised my eyes to the sky and could see nothing but the worthless timber that covered everything," wrote another. Even if you succeeded in knocking one of these monsters over, how would you dispose of it, much less remove the sprawling stump so you could do something useful with the land like plant crops or feed your animals? Some advocated abandoning the region altogether; as late as 1881, when settlers had already established a solid foothold on the North-west Coast, a London magazine editor wrote, "British Columbia is a barren, cold mountain country that is not worth keeping. Fifty railroads would not galvanize it into prosperity." Prosperity, of course, was the name of the game; the Bible decreed it, and the government encouraged it. If not for profit, advancement, or adventure, why else would one leave all that was safe and familiar to do battle with giants? The notion of forest conservation, a practice that had only recently caught on in Europe, was anathema in a land of such daunting bounty. The problem of the day was not how to preserve or manage the forest, but how to master it, fulfill the mandate of manifest destiny, and turn this infinity of trees, and the land on which they stood, into something *productive*.

IN 1852, FAR TO THE SOUTH, the first Giant sequoia was felled—not for the fantastic amount of wood it contained—but simply to prove that it could be done. However, with the California gold rush in full swing and San Francisco booming, it didn't take the Americans long to figure out what to do with all that wood. Within a decade, they had secured a virtual monopoly on the West Coast timber market. Companies with names like the Douglas Fir Exploitation and Export Company were doing a brisk business out of San Francisco, handling the wide and flawless timber coming south from the coastal sawmills of Oregon and Washington. Meanwhile, north of the border, in British Columbia, a wood supply that dwarfed even the vast U.S. reserves was languishing by comparison. As early as 1864, the *British Columbian* lamented that

the numerous and extensive milling establishments on Puget Sound [Washington] have enabled our enterprising neighbours . . . to enjoy much of a monopoly of the great lumber trade of this Coast. Although we have harbours and pineries not one whit inferior to theirs . . . they, having so much the start of us, have thoroughly established trade, whereas we have to a great extent yet to make ourselves known abroad.

Canada was not yet confederated when this was written, but it articulated a disadvantage that, to this day, continues to plague the country, which has a population and a GDP one tenth the size of its southern neighbor's. In an effort to rectify the situation, resource maps and promotional pamphlets with titles like *British Columbia's Supreme Advantage in Climate, Resources, Beauty and Life* were liberally distributed in the east. "It makes little difference to the people of western Canada where the money comes from," observed a turn-of-the-century writer in a trade magazine called *Western Canada Lumberman*, "as long as the country is developed." In keeping with the spirit of the times, Vancouver's official motto pays no Latin lip service to Truth, Duty, Faith, or Light; instead, it sounds more like a corporate slogan: "By Sea and Land We Prosper." Not surprisingly, a lot of development capital came from American investors. John D. Rockefeller optioned thousands of acres of prime forest on Vancouver Island, while Michigan timber magnate Frederick Weyerhaeuser, along with the famous California railroad owner and university founder Leland Stanford and others invested in railroads whose primary purpose was to access lucrative B.C. timberland.

Technical expertise was imported as well; it was Matt Hemmingsen, a logger out of Wisconsin, who was brought out to Vancouver Island to break up one of the biggest logjams in West Coast history. Most of the early loggers on the coast were easterners coming out of Nova Scotia, Maine, and the Midwest, where floating logs downriver to market was standard practice, but the huge timber of the Northwest was ill-suited to this method as it tended to run aground. A particularly bad logjam could pile up as much as eighty feet high, and when Hemmingsen arrived on the scene, he was confronted with a tangled snake of giant timber six miles long. In the end, he broke it up by blasting all the river bends.

British Columbia's timber industry didn't really come into its own until after World War I, and it was due in large part to Harvey Reginald MacMillan. "H.R." MacMillan was a penniless, fatherless boy from Ontario who entered Yale's school of forestry in 1906; he went on to become B.C.'s first chief forester and, later, a bona fide timber tycoon who, it was said, "would be selling to the moon if he could get delivery." He very nearly did; in 1915, in an effort to challenge the U.S. timber industry's stranglehold on West Coast exports, MacMillan literally circumnavigated the globe, drumming up business for B.C. wood products. His efforts paid off handsomely, and for much of the twentieth century his name was synonymous with Canada's largest wood products corporation. In time, MacMillan Bloedel's holdings would extend from southeast Asia all the way to the Yakoun River and the golden spruce.

CHAPTER FIVE

The Beginning of the End

Fancy cutting down all those beautiful trees   to make pulp for those
bloody newspapers, and calling it civilisation.

—Winston Churchill, remarking to his son during a visit to Canada in 1929

I T IS HARD, even now, to imagine the magni-
tude of the timber coming out of the Northwest forests throughout
the nineteenth and twentieth centuries. Photographs taken anywhere
along the coast, from southeast Alaska to Northern California, show
sturdy men in heavy clothing dwarfed against backdrops of mono-
lithic cylinders so large that they are scarcely recognizable as trees.
They look, instead, like oddly symmetrical boulders, or the fallen
columns of gargantuan temples, which may be closer to the truth. An
elderly Haida man who spent much of his life in the Yakoun Valley
falling trees for a southern lumber company indicated the breadth of
the logs he dealt with every day by glancing at his ceiling. "You'd

gouge into the ground that deep, too," he explained. "You'd be covered in mud from head to toe."

The work of removing rainforest timber is not only wet to the point of amphibiousness; it is also hazardous in the extreme. Even today, despite the advent of elaborate safety regulations and state-of-the-art equipment, the odds that a logger will be killed on the job are approximately thirty times greater than those of the average North American worker. Cutting the tree down is only one high step in a long and arduous journey that begins with gaining access—for man, beast, or machine—to an often trackless wilderness, and ends with delivery to the marketplace that may lie a continent away. The act of felling a tree is, relatively speaking, the briefest of events; it is to the timber business what the act of conception is to raising a family: a beginning that really happens somewhere in the middle. And both acts capture the imagination for similarly cataclysmic reasons; they are thunderous and defining moments, after which nothing is certain except the knowledge that the hardest labor lies ahead. A single large log, bucked to thirty feet and limbed for transport, could still weigh fifty tons, as much as a loaded semi; an entire tree might weigh five times that. Somehow this massive object—half steamroller, half battering ram, and slippery as an eel—has to be moved out of the forest, which may be growing on a mountainside with a 45-degree slope. In some places, like the Yakoun Valley, they had to resort to methods normally reserved for quarrying stone: many logs were so big that they had to be riven—split lengthwise—with "dynamite wedges" in order to be moved at all.

IN THE EARLY TWENTIETH CENTURY, a single logger, equipped with a government hand-logging license, an ax, a saw, and a Gilchrist jack, could roam the coast more or less at will and ply his trade by dropping favorably positioned trees directly into the salt-chuck—the sea. Gordon Gibson, who is legendary among old-time West Coast

loggers, started out as a hand logger before becoming a major operator and politician on the coast of British Columbia. In 1933, while
searching the now-famous Clayoquot Sound area of Vancouver Island
for promising trees, he found a particularly memorable one a thousand feet up a mountainside. It was a Douglas fir —to this day, the
preeminent commercial species on the Northwest Coast—and it was
a textbook specimen. 14 feet in diameter and 225 feet tall, its perfectly cylindrical trunk ascending for 100 feet before the first branch
broke the symmetry. Gibson and his men started into it with a crosscut saw, the theory being that if they felled the tree downhill, it
would make the thousand-foot journey to the water under its own
terrific momentum.

However, before any cutting could begin on a tree like this, a
series of steps would have to be built into the trunk in order to get
above the tree's broad base, or butt-swell, where the roots began to
fan out. Since a big tree's butt swell might begin well above a man's
head, a faller would cut a series of notches into the side of the tree,
about six inches deep and the size of a mail slot. Once the cuts were
made, a sturdy plank, called a springboard, would be inserted end-
wise into each notch. This crude portable scaffolding is what the fallers would stand on as they chopped and sawed their way through big
West Coast trees. A metal "shoe" with a sharp lip kept the springboards from slipping out as they bounced up and down with the
rhythm of the men's work. Neil McKay, an old-timer from Vancouver
Island who came into the business when horses were still a common
feature in the woods, recalls springboards reaching five and six levels
high. They are still used on occasion today.

Despite changes in equipment, the technique of bringing a tree
down has changed little over time. The goal has always been to bring
a vertical shaft into the horizontal in as controlled a fashion as possible—and with minimal damage to the tree. (Some fallers would
make beds of branches in order to prevent the trunk from shattering
on impact, but this is almost impossible to do on a mountainside.) A

controlled fall is accomplished by creating a hinge at the point of cutting. After determining the optimal falling direction—usually a three-way compromise between personal preference, the tree's natural lean, and the lay of the land—a "wedge" is cut facing in the desired direction of fall. If a saw is available, a horizontal "undercut" is made about a third of the way into the trunk—a serious undertaking, in the case of a big West Coast tree. Prior to the chain saw, this would have been accomplished in stages—a cut would be made and then the wood above would be chopped out with an ax to make more room and reduce the friction imposed on the saw blade. Then another cut would be made, followed by more chopping, and in this way the fallers would cut and hack their way into the tree. Many early photos show a whiskey bottle hanging from a tree within arm's reach of the fallers; these bottles were filled not with whiskey to lubricate the men, but with oil to lubricate the saws as they worked their way into the wet, often sappy trunks. Gibson's crosscut saws—broad ribbons of toothed steel with a broom-style handle at each end—were six and ten feet long, and his double-bitted axes were more than a foot wide from blade edge to blade edge. With two men working steadily, perched on opposing springboards, Gibson's fourteen-foot, 800-year-old "Doug" fir would have taken all day to bring down. At dusk, when the heartwood finally gave way with a sternum-shuddering groan, the men dropped their tools, jumped from their perches, and fled uphill into the thick salal that covered the steeply angled forest floor. From there, they watched as the fruit of their labors—weighing about as much as a jumbo jet—came crashing down to earth. Wrote Gibson:

> It seemed to pause in the air for a moment like an eagle in
> slow motion before starting down the mountainside, cart-
> wheeling, end over end and disappearing into the water at a
> 45-degree angle. After what seemed to be a five-minute lapse,
> it suddenly emerged on the surface like a giant whale breach-
> ing from the depths. It was completely devoid of branches and

most of its bark had been stripped away by the 1000-foot passage over rocks and windfalls.

In his memoir *Bull of the Woods*, Gibson neglects to describe the sound a tree that size would have made as it tumbled down the mountain; it would have been absolutely thunderous—an echoing, earthshaking avalanche of one. Old-growth West Coast trees are the heaviest objects routinely dropped anywhere in the world.

For reasons like this, West Coast logging, especially in its early twentieth-century heyday, was not so much tree-cutting as it was a kind of terrestrial whaling: determined, poorly paid men working in remote areas were using temperamental machinery and simple hand tools to subdue enormous, often unpredictable creatures that could squash them like bugs—and did. One county on the Washington coast lost more than a hundred men to logging accidents in a single year (1925), and there are—still—lots of ways to die.

In terms of fatalities per capita, commercial bush pilots have the most dangerous job in North America, but theirs is a small niche profession. The number two position is shared by the far larger populations of commercial fishermen and loggers, but comparing the two is misleading. Fishermen are frequently killed en masse when their boats go down, while loggers are almost always killed one at a time. Thus, in terms of the sheer number of accidents, loggers have much more in common with bush pilots. When one considers the fact that loggers work on land during daylight hours, often with winters off— as opposed to at sea or in the air around the clock and in all kinds of weather—it becomes clearer just how treacherous logging can be. And loggers have everyone beat when it comes to variety: pilots generally crash and fishermen generally drown, but loggers are killed and maimed in a chilling assortment of ways that combine aspects of industrial accidents, warfare, and torture.

Frank Garnett was a turn-of-the-century settler and ox team logger who had made a name for himself by felling some of the biggest timber on Vancouver Island, a place well known for its enormous trees. When gravity wasn't available to carry trees to the water, oxen and, later, horses were the alternative. After a tree had been bucked up (sawn) into lengths, the log ends would be "sniped"—tapered with an ax so they would slide more easily across the rough ground; they were then linked together with heavy hooks and chain. Often the train of logs would be taken out over "corduroy" roads, also known as skid roads, made of smaller logs laid crosswise across the path (the latter is the origin of "skid row"). To ease their passage, the causeway would be lubricated with water, crude oil, whale oil, or even dogfish oil, which was rendered from a kind of shark once plentiful along the coast. Once the logs were in line, the ox team, consisting of perhaps a dozen animals, would be harnessed to the lead log; in order to keep from slipping on the oiled roadway, oxen and horses would be shod like their human counterparts: with spiked shoes. When all was ready, the bull whacker, using that astonishing combination of tender endearment and paint-peeling invective unique to drivers of animals and machines, would cajole his team into motion.

Garnett was driving his heavily laden ox team out of a stand he'd been working in Maple Bay when one of the logs shifted, pinning him between another similarly massive log. The log rolled in such a way that neither man nor beast could move it, leaving Garnett trapped but very much alive. His mother happened to be on hand, and while attempting to comfort him, she found herself in a bind nearly as hideous as her son's. Garnett—in agony and knowing he was doomed—begged his mother repeatedly to put him out of his misery with a nearby sledgehammer. Unable to kill her own child, even in an act of mercy, she endured his pleas for two hours until he finally bled to death.

A contemporary of Garnett's was Freeman Tingley, an American who was one of the first pioneers to homestead on the Queen Charlottes. Tingley helped to settle the future logging town of Port Clements, not far from the golden spruce, and he was one of the few early arrivals who wasn't starved out after a year or two. In addition to hand-logging with oxen well into his seventies, Tingley was known for growing enormous vegetables which he sold to natives and settlers alike. Successful on other fronts as well, he was nicknamed "Stud" in recognition of his multiple wives and copious offspring. One of Tingley's many grandsons, Harry, was born on the islands in 1928; at that time, the sound of Haida drumming could still be heard at night, sometimes pulsing across Masset Inlet, near the terminus of the Yakoun River. Harry Tingley began a lifetime in the logging industry at the age of fourteen. He was entering a world that is hard to imagine today, a place where a boy's first day on the job might begin in a mud-filled ditch with a knee on his chest and a filthy hand scraping the peach fuzz off his face with a hunting knife. "How'dya like that?" his new friend might say as he stumped off. "Now yer a man."

Tingley retired in 1993 at age sixty-five, but statistically speaking, he should probably have died long before that. Tingley lost one brother and two half brothers to logging accidents, and that was just the beginning. Like veteran bush pilots, career loggers can rattle off the names of dead and crippled comrades in numbers whose only comparison can be found among professional soldiers. One friend of Tingley's named Vaillancourt was scalped by a flying cable; another named Judd McMann got dragged through an edger—a machine used for squaring up the edges of rough boards. "Mighty" Joe Young broke his back when he was thrown from a skidder after losing control while hauling logs off a mountainside. Carl Larsen was hit so hard by a cable that the impact tore the aorta out of his heart; another Swede Tingley knew was killed when the tree he was felling kicked back, striking him from behind as he ran. Still another managed to make it safely back to Vancouver only to lose his life when he stepped out of a Granville Street bar and was shot through the head by a drunk firing at random from a block away. For years, Vancouver's *News Herald* maintained a loggers' death count box, just as the *New York Times* does for American soldiers during wartime.

Tingley, who has logged Queen Charlottes spruce up to sixteen feet in diameter, seems to have survived on luck and a cat's reflexes (another half brother fought three rounds with Jack Dempsey), but raw optimism may have played a role as well. Describing one incident in which he dove into a shallow ditch to soften the blow of a runaway log, he said, "You know, the body *will* compress a little when a log rolls over it." But sometimes Tingley was too quick for his own good. Once, while on a drunken tear, he made the mistake of insulting a big Italian named Furdano. The man took exception, knocked Tingley to the ground, and proceeded to kick his head in; if another logger named Bear hadn't intervened, Furdano probably would have succeeded. Tingley was carried, unconscious, to his bunkhouse and put to bed, where he was awakened the next morning by the foreman—the "push"—coming in to fire him. Tingley could only hear the man's voice because he was blind; he had been kicked so hard that his eyes had filled with blood. When the push saw Tingley's face he gasped and rushed out of the room; Tingley overheard him in the hallway telling the first-aid man, "His eyes are gone; he's lost his eyes."

Tingley recovered, but it was beatings like this that spawned the term "logger's smallpox" – a reference to the scars left behind after a stomping with calk boots. Pronounced "corks," these boots resemble high-top industrial golf shoes: their soles have three-inch heels and are studded with half-inch spikes; the uppers are made of heavy, tasseled leather and they lace up to the calf—sometimes to the knee. In a forest of slippery wood and loose, moss-covered rock, they are as necessary as crampons are for an alpinist.

Sometimes the forest would simply swallow a man whole. In the early 1960s, at a camp outside Jeune Landing on Vancouver Island, a crew was working a patch of blowdown—forest that had been knocked over by a windstorm. Many of the trees had come up by the roots and the men were cutting the fallen trees off at the base, leaving the stump and root ball—some of which were twenty feet across—standing up on edge. At midday the crew broke for lunch, but when they reconvened, two men were missing. Search parties were brought in, but to no avail; it was as if the men had simply vanished off the face of the earth. It was only when all other possibilities had been exhausted that it occurred to someone to look underneath the stumps, and that is where they found the missing men. They had made the mistake of using the shady underside of one of the upended root balls as a backrest, and while they ate, the stump had fallen back into place, closing over them like a great earthen jaw.

Accidents were so common in the early days that, if a man was killed on the job his body would simply be laid to the side and work would continue until quitting time, when a boat, plane, or runner might be sent to notify the police. In remote areas, this practice continued at least into the 1980s. Even now, loggers must sometimes carry their dead partners out of the woods like sacks of flour. Many camp foremen saw their workers as expendable, interchangeable units to be hired and fired at will; there were camps that were known for having three crews: one on the job, one that had just been fired, and another coming in on the next boat. Panicky Bell was a notorious foreman on the Charlottes, and Harry Tingley recalled an incident in which two men had been fired. Bell called for a plane to take the men out, and when the pilot balked at coming all that way for just two men, he fired several more on the spot, just to round out the load.

A foreman's reputation (and pay) was enhanced or diminished based on the productivity of his crews, and the result was a practice known as highballing. Highballing entailed hauling logs out of the woods as fast as humanly and mechanically possible—no matter what. Harry Tingley calls it "working like Billy Be-Damned." Once horse and oxen were replaced by steam donkeys —steam-powered winches that would haul logs out of the woods with cable and pulleys—the pace accelerated exponentially. When donkey engines first came on the scene at the turn of the century, they were a simple, relatively small apparatus not much bigger than a Dumpster. They rode on sleds made from logs and they would move themselves through the bush by being cabled up to a distant tree and then reeling themselves in. The first donkeys were designed to drag the logs out along the ground, but as technology improved, the high lead and skyline were introduced. A high-lead cable was run from the donkey to a tall stout tree (the spar tree) high on the mountainside, while skylines ran between two spar trees. Secondary cables called chokers were suspended from the high lead and attached, one apiece, to the end of a log; this enabled loggers to haul clusters of logs out in a partially airborne state, thus avoiding all the hang-ups on rocks and windfalls that ground lines ran into. Where an ox team would move a load at a slow walk, a steam donkey with a high-lead cable could haul a cluster, or "turn," of logs out of the woods at 30 miles per hour. A turn of thirty-foot logs bouncing at that speed over rough ground takes on the characteristics of what one eyewitness described as a crazed fifty-ton kangaroo. When accidents happened they tended toward the catastrophic.

It is always amazing to consider the things a man will risk his life for, and the promise of a case of beer on Friday could mean the difference between an average week of log production and a record-setting one; it could also mean the difference between life and death. But as the machines got bigger, more powerful, and more expensive to own and operate, the expectations only rose. Neil McKay remembers an enormous machine from the 1920s called the Washington Flyer; it was a steam donkey eighteen feet long by eleven feet wide; it rode on a ninety-foot sled made from logs that were five feet in diameter, and it could work 3,500 feet of two-inch skyline cable. "It was a monstrous goddamn thing," recalls McKay; "you could clear a whole mountainside with it." And so they did, the gear running so hard that pulleys and cables sometimes glowed red hot and set fire to the surrounding woods.

Since the donkey puncher (the man operating the donkey) might be half a mile or more away from the chokermen who were "setting" the chokers around the logs ready to be yarded (hauled) out of the bush, a code of whistles was used to communicate what was needed down the line: haul back; ease forward; stop, etc. There was even a special death signal: seven long blasts. The man in charge of passing this information back and forth was known as the whistlepunk and he would be stationed on a tree stump or some other piece of high ground where he could see the big picture. Below him, looking like so many squirrels in the face of these huge trees, mountains, and machines, the fallers and chokermen spent their often rainy days beset by swarms of mosquitoes and biting flies as they scrambled over giant logs, rocks, and the sharpened points of broken branches, struggling to meet or surpass their foreman's quota. It was an environment that seemed scientifically designed to crush limbs and puncture bodies, and it is no wonder that men were frequently injured or killed, or that they often quit, or that they might eat three T-bone steaks, a plate full of potatoes, and a serving bowl full of ice cream for dinner.

By the mid-sixties, the days of the steam donkey were a decade past, and the look, feel, and sound of logging had changed forever. The soft *chuff-chuff-chuff* of the wood-fired boiler had been replaced by the clanking roar of the diesel engine, and the era of truck logging, the kind most people see today, was in the ascendant. The last axmen had gone over to chain saws by the early fifties—though these early power saws weren't much of an improvement. In addition to being mechanically cranky, the steel and magnesium machines had seven-foot blades and could weigh 140 pounds. Tree falling fifty years ago would have been kind of like climbing mountains all day while carrying a Harley-Davidson motorcycle engine, along with its chain and rear sprocket. By around ten in the morning, a five-pound ax must have looked pretty attractive. However, the power and appeal of the chain saw is undeniable, and even then it was clear they were in the woods to stay. But no one was prepared for the impact they would have on the forest.

CHAPTER SIX

A Boardwalk to Mars

Next year we will practice havoc
in that green trench—
the saws will yammer their nagging dirge,
the donkeys will gather the corpses,
the land will be hammered to stumps and ruin.
— Peter Trower, "The Ridge Trees"

ANGUS MONK WENT into the woods at the age
of thirteen. This wasn't so unusual in the 1920s when millions of
boys, rendered fatherless by World War I, were forced by circumstances to leave home and make their way in the world. On the West
Coast in those days, finding a boy as young as ten taking a shift at the
wheel of a tugboat while the skipper slept, or paddling a dugout
canoe to an island that might lie several miles offshore, wasn't out of
the ordinary. It was a different time: the country was huge, the population small, and the workload enormous; competence—in any

form—was exploited to its fullest extent. The Monk family had come to Canada from the tiny Scottish island of Benbecula in the Outer Hebrides; this barren, windswept archipelago is Europe's answer to the Queen Charlottes—minus the trees. The rigors of life out there, at the mercy of the North Atlantic, necessitated a tough and fearless constitution, a characteristic that Angus embodied fully. Naturally drawn to outer limits, both physical and geographic, he made his way to Vancouver Island, where the logging camps served as his high school and university. Though barely pubescent when he started, Angus proved an apt, if restless, pupil. Drifting from camp to camp, he picked up a wide variety of skills, emerging as an exceptional high rigger, the most dangerous and highly paid job in the woods.

If you run across an old photo of a man with an ax perched high in a tree, you are not looking at a faller, you are looking at the far rarer high rigger. Like the steelworkers who build skyscrapers, these men were a breed apart; they were the ones who prepared the spar trees to run high-lead cable. A high rigger's duties included hanging the huge pulleys—three feet across and two thousand pounds—that carried the cable, and setting the guy lines that anchored the spar tree from all sides in order to keep it from being pulled over by the tremendous loads it would be supporting. It was a job that required an unusual combination of raw courage, gymnastic strength, and technical skill; the success of a logging operation hung, literally, on the competence of its high rigger.

The first high riggers were sailors; already at ease at the tops of tossing masts and familiar with complex rigging, they were naturals for the job. A high rigger's special equipment was minimal: three-inch climbing spikes which were strapped to his ankles, and a heavy rope that was clipped to his waist and slung around the tree's trunk. The rope was woven with a wire core in order to prevent it from being accidentally cut by the rigger's ax—a shorter, more compact version of those used by fallers. Also attached to his belt, in addition to the ax and a one-man crosscut saw, would be a long "straw line"; this light rope was used to haul up successively larger ropes, cables, and blocks (pulleys) once the spar had been made ready. Thus prepared, a high rigger like Monk would "hug" his way up trees 250 feet high, lopping branches as he went. Because the topmost portion of the tree was thinner and not as strong, it would be cut off; sometimes it was blown off with dynamite. Topping the spar was a tricky operation; if a breeze came up and the top began to fall before it had been cut through completely, the tree could "barber chair"—split open along its length—with the result being that the rigger would be crushed between the expanding tree and his fixed safety rope. For trees this tall to survive winter gales, they must be extremely flexible, so even when everything went according to plan, the release of the several-ton spar top would cause the tree to sway wildly. Coworkers on the ground would watch as the tiny rigger ducked his head, dug in his spikes, and held on for dear life while the spar thrashed back and forth like a ship's mast in a storm. After all was calm again, some high riggers, including Angus, would stand on the platform they had made—about as wide as a cocktail tray—and urinate into space. Once all the blocks were in place, the rigger could then rappel down; Angus became so adept at high-speed descents that he could throw his hat in the air at 150 feet and be on the ground when it landed. For these reasons, among others, high riggers were viewed with a certain awe by other loggers—a combination of admiration and relief that the job wasn't theirs. Once back on earth, such a man was entitled to a certain swagger; part stuntman, part matador, and absolutely indispensable, there was no doubt he was a card-carrying "cock of the woods."

But even high riggers were employees, and Angus had higher aspirations. Eventually he learned enough on the job to go out on his own, becoming what was known in the industry as a gyppo logger; it was a big and precarious leap. Gyppos were independent operators who might own a few trucks, a couple of mobile camps, and the shows they ran tended to reflect their own characters, for better or worse. Like their agricultural equivalent, the freehold farmer with sixty cows and a quarter-mile section, they were extremely vulnerable to the whims of the marketplace, and they are an endangered species today. During the late fifties and sixties, Angus and his crew were clear-cutting the valleys above Howe Sound, a deep fjord that feeds into Vancouver's English Bay from the north. Though often shrouded in clouds and fog, it is a stunning setting of deep shining water dotted with islands and overseen by mountains cloaked in forest. Winding through West Vancouver and up the sound's east side is the Sea to Sky Highway; the road, which went through in 1958, was an engineering feat on a par with California's Highway One. As grand entrances to cities go, it rivals San Francisco's Golden Gate; there is no other highway on the continent that places a traveler so tightly between the jaws of mountain and sea. Angus Monk had a contract to log the steep slopes above it.

VANCOUVER IN THE 1950S was still very much a British colonial town, and government, social mores, and education all reflected this. Cut off from the rest of Canada by the Rocky Mountains and the Coast Range, and from the United States by the border, Vancouver, with the mildest climate in mainland Canada, floated in a lush green world all its own. To this day, the west side still has the feel of a British colonial suburb: like those in Cape Town, Hong Kong, or Penang—only with English weather. When it isn't raining, sailboats cruise the bay beneath snowcapped mountains while cricket, lawn bowling, and tennis are played at the clubs ashore; "Royal" and "British" are common prefixes. Fig trees, windmill palms, and Japanese bananas flourish alongside Chilean monkey puzzles and tree-sized camellias in a climate that feels closer to California than to Canada.

Vancouver's east side was a separate world from the serene and stately west side suburbs. In the dense shingle and clapboard neighborhoods that rose above the docks and lumber mills, immigrants from Asia and Europe jockeyed for position alongside Indians from all over western Canada. As spread out and isolated as loggers might be in the bush, the downtown east side is where they all came together. Here, transient loggers, miners, and fishermen drank in gender-segregated bars on downtown Granville Street, and whored to unconsciousness in flophouse hotels like the Blackstone and the Austin.

For generations, loggers have been viewed as a kind of subspecies that require special handling, like boxers or British football fans. In one telling instance, the radio operator aboard the M.V. *Princess Maquinna*, a passenger steamer serving the B.C. coast, called the port at Vancouver to say, "We've got fifty passengers, and one hundred and fifty loggers." In many cases, derogatory nicknames like "bush ape" and "timber beast" were more than justified; as one man in a position to know put it, "There were some awful bloody animals in the woods in those days." For a certain segment of the population, logging was B.C.'s answer to the French Foreign Legion: junkies, petty criminals, and thugs would often seek refuge in the camps, and the courts encouraged it. But even way out there, heroin was available.

Coming out of the bush by boat and plane, loggers were walking liabilities; not only were these men in formidable physical condition, they had vicious cases of cabin fever and were desperately oversexed. Many would already have a bottle in hand as they headed to town primed to blow off a serious head of steam. For Bill Weber, a forty-five-year-old faller from Vancouver Island, those days are recent memories. Born in a tiny logging community on Vancouver Island, Weber's father is a preacher who made his living not off the collection plate, but by driving logging equipment between sermons. His grandmother was one of the last children to travel west in a covered wagon. Weber stands six-foot-three in his calks and with his Bunyanesque physique, piercing blue eyes, and flaxen hair he looks like a Teutonic knight, or a poster boy for the logging industry. He recalls one seaplane flight where, already well fortified, he decided that he needed to relieve himself immediately. Much to the pilot's consternation, Weber opened the door and climbed out onto the plane's pontoon and into a 100-mile-an-hour headwind. With one hand on a wing strut and the other on his fly, five fingers were all that kept him from a thousand-foot plunge into Georgia Strait. After several months in the bush, a man might have a lot of money coming to him, and the wads were fat and heady. "I'd have three or four thousand dollars in my shirt pocket," recalls Weber, "and I'd be struttin' around like I had the world by the tail on a downhill pull."

Most young loggers were strangers to city ways and they made easy marks; it was for reasons like this that two downtown Vancouver side streets earned the names Trounce Alley and Blood Alley; a third is called Shanghai. They still exist today. While there were also stories of Indians who would roll drunken whites and leave them draped across the train tracks outside the freight yards, Vancouver had one advantage in that it is the only city in mainland Canada where you can pass out in a park on a winter night without freezing to death. "I'd go to Van, blow my paycheck, and fly back out with the clothes on my back," recalled Weber. "There was a lot of booze and dope, snoose, Irish coffee in the thermos—it was a pretty integral part of the life. If a guy was half drunk, the crew would cover for him."

They would do so not only because they would expect the same in return, but because their lives depended on it. Even today, it is not uncommon to see a logger covering for an ailing partner, or to find a man on the mend simultaneously chewing tobacco and smoking a cigarette in order to settle frayed nerves or an upset stomach. There is little doubt that drugs and alcohol have played a role in a number of deaths; one Vancouver Island powder man returned to the woods so snaky (addled by DTs) that after setting a fifty-pound charge under a big stump, he then sat down on it and blew himself up. "Never take me off the ground on a Sunday," Angus Monk would say, referring to the hangover he would inevitably have. Angus was not so different from other loggers of his generation for whom alcohol was, for all practical purposes, one of the basic food groups, but he took it to some unusual extremes. Harry Purney, an old friend from the age of steam, recalls him preparing the following concoction one morning and calling it breakfast:

EGGS À LA ANGUS

Boil and peel 17 eggs
Place eggs in a bowl
Add a cup of Cutty Sark
Serves one

Angus, who clearly had appetites as formidable as his constitution, was nonetheless able to achieve some sort of balance, albeit a precarious one; by mid-century loggers' standards, he had the best of both worlds. Whereas most woodsmen of his era were banished for months at a time to remote valleys accessible only by boat or float plane, Angus ran his crew in the bush all day and then got to drive home to his family in the exclusive suburb of West Vancouver. He was a tough, happy, connected man, remembered with fondness by all but his rivals, the only thing missing was a son to work beside him. For a brief time, a nephew would fill the gap. Angus's sister, Lillian, had two boys, but only one of them would make it past thirty five; his name was Grant Hadwin and there were ways in which he could have been Angus's own. Both uncle and nephew shared a frontiersman's capacity for excess and risk. In 1966 Grant quit school and left home; he was sixteen, and his first boss was his Uncle Angus.

Logging is a brutal trade that can make one an old man at fifty, but Grant, a natural athlete, was ideally suited to the task and excelled in the face of its physical challenge and myriad dangers. The rugged, isolated lifestyle, a last link to the rapidly vanishing frontier era, captured his imagination and would shape his life, not least because it flew in the face of his engineer-father's white-collar ambitions. By the time he was twenty, Grant Hadwin, an upper-middle-class prep school refugee, had adopted the costume and habits of the old-time loggers: gray wool Stanfields; stagged* jeans held up with suspenders, a lower lip full of Copenhagen, and an alarming capacity for alcohol.

*Jeans cut off just below the knee to prevent snagging.

As dramatic as the transformation was, it was less a reaction than the natural evolution of a person finally allowed to follow his organic impulses. During high school, while his peers were racing cars and chasing girls, Grant was building a cabin and roaming the mountains that reared up behind his parents' West Vancouver home. Few people had the chance to get to know Grant very well, in part because he seemed to be in constant motion; one of his only high school friends died young in a motorcycle accident. Hadwin is remembered by other classmates as being a loner, and very independent-minded. "He was really intense," recalled Truls Skogland, who knew Grant as a fifteen-year-old, "not negative—he just had spirit." Skogland and others recalled him as a gifted tennis and rugby player and "a wizard on the pegboard."* Despite an apparent preference for mountains over people, Grant was disarmingly courteous and well spoken. His paternal aunt, Barbara Johnson, recalls her young nephew having "the most beautiful manners. He was quite self-assured, and so polite and nice." If good manners and comportment were the measures of a successful education, Grant's early years in boarding school had certainly paid off: "He was very polished," recalled another classmate; "he could have had an audience with the queen."

Grant's father, Tom Hadwin, had graduated at the top of his class from the University of British Columbia's electrical engineering program, and he went on to become a top engineer and lifelong employee at BC Hydro, the province's biggest power company. "You never won an argument with Tom Hadwin," recalled one employee. Tom Hadwin was a very different kind of man from Angus Monk; where Tom was tight, degree-conscious, and cerebral, Angus was expansive, irreverent, and vigorously physical. Grant loved him; compared to the stifling atmosphere of home and school, Angus was pure oxygen. But even as he idolized his uncle and the life he represented, Grant was horrified by what he saw. From the very beginning, the Faustian

---

*A gymnasium exercise in which one climbs a vertically mounted board by inserting handheld pegs into successively higher holes.

bargain most loggers are forced to make wasn't lost on him. After an early stint working with his uncle, Grant returned, briefly, to West Vancouver, where he visited his Aunt Barbara—one of the few family members with whom he stayed in touch. According to her, Grant was struck by the destructiveness of the logging process. Then only seventeen, he described logging techniques that stripped the mountainsides down to bare rock. "Nothing's going to grow there again," he told her.

This was an unusual thing for a teenager from Vancouver to be concerned about in 1967, especially one with Hadwin's lineage. Logging had literally built the city and most people were still connected to the industry—if not directly, then through family members or friends. But things were changing in the sleepy green logging town—quietly at first. Not long after Grant had reported his observations to his aunt on the north side of English Bay, a fledgling organization formed on the south side, just five miles away. They gave themselves a deceptively Canadian name, the Don't Make a Wave Committee, but this would prove a misnomer, and in 1970 they would change it—to Greenpeace.

BY THE TIME Hadwin got into it, the West Coast logging industry was highly mechanized, but still almost completely unaware of environmental issues; there was only sporadic replanting of logged areas, and conservation, as we know it today, was of negligible concern. On both sides of the border the forests of the Northwest were still being treated like a kind of inexhaustible golden goose. The relationship between local governments and the timber industry was generally one of self-serving collusion with the emphasis being on volume and speed; the working motto was "Get the cut out." It was nothing to clear-cut both sides of an entire valley and simply move on to the next; in fact, it was standard procedure—decade after decade, and valley after valley. After all, there were so many, especially up in British Columbia.

By any measure, B.C. is an absolutely enormous place; it occupies two time zones and is bigger than 164 of the world's countries. All of California, Oregon, and Washington could fit inside it with room left over for most of New England. From end to end and side to side, the province is composed almost entirely of mountain ranges that are thickly wooded from valley bottom to tree line. Even today, it is hard country to navigate; the drive from Vancouver, in the southwest corner, to Prince Rupert, halfway up the coast, takes twenty-four hours —weather permitting. Though the province is the width of Texas, there are only two paved roads accessing its northern border, and one of them is the Alaska Highway. B.C.'s coastline—including islands and inlets—is 17,000 miles long, and all of it was once forested, in most cases, down to the waterline.

Like Alaska, this landscape exudes an overwhelming power to diminish all who move across it. A colony of thousand-pound sea lions might as well be a cluster of maggots, and a human being nothing but an animated pouch of plasma for feeding mosquitoes. That something as small as a man could have any impact on such a place seems almost laughable. In a geography of this magnitude, one can imagine how it might have been possible to believe that the West Coast bonanza would never end. And the numbers bear this out; B.C.'s timber holdings were truly awe-inspiring: in 1921, after more than sixty years of industrial logging, an estimate of the province's remaining timber came in at 366 billion board feet—enough wood to build twenty million homes, or a boardwalk to Mars.

HADWIN, TRUE TO FORM, didn't stay with his uncle long. After a brief apprenticeship, he headed deep into the Coast Mountains to the former mining town of Gold Bridge, four hours north of Vancouver. Hadwin already knew the area well: his family had owned a cabin on Big Gun Lake, just outside of town, since he was a child. Insulated from the outside world by a natural fortress of high, rugged mountains, Gold Bridge has always been a marginal place. The rivers there run glacier green, and the only access is via rough logging roads lined with fatal drop-offs. A few miles to the south lean the ruins of the Bralorne-Pioneer mine, the most lucrative gold mine in British Columbia. In its heyday it employed thousands of men who worked as much as a mile underground, breathing dank, recycled air that was in excess of 120 degrees. When the mine closed in 1971, it caused the local population to plummet to fewer than one hundred souls. Today, grizzly bears, wolves, and mountain goats outnumber people.

Before finding his calling as a commercial timber scout and layout engineer, Hadwin worked variously as a logger, prospector, heavy-equipment operator, blaster, and hard-rock driller. Between jobs, he spent a lot of time on his own, hunting and exploring the surrounding wilderness. In the evenings he seemed to oscillate between the two poles represented by his father and his uncle: the surprisingly suburban pastime of contract bridge, and rowdy nights in the local bars. One of Hadwin's neighbors recalled him and a man named Franklin stretching their penises across a bar table in order to determine whose was longer while a native woman named Big Edith refereed. No doubt similar contests went unrecorded in Dodge City a century earlier.

On another occasion, a man in the Bralorne bar bet Hadwin a hundred dollars he couldn't climb a thousand vertical feet in an hour. This is no mean feat in the Coast Range, which is composed of steep, slab-faced mountains where 40- and 50-degree slopes covered in loose rock and snow are commonplace, but Hadwin left the bar and returned shortly with the money. When asked where *his* money was, the man, realizing with whom he was dealing, reneged. But Hadwin went out and timed himself anyway, just to prove that he could do it. Hadwin was a person who threw himself fully into every undertaking; his stamina and competitiveness were the stuff of local legend and he was well known for running his coworkers into the ground. "Even with his hands in his pockets, he could scamper over stuff like you wouldn't believe," a former assistant, who now works for the

Ministry of Forests, recalled. "You needed a jet pack to keep up with him."

"He was in the best condition of any man I've ever seen," explained Paul Bernier, a longtime colleague and close friend. "We'd run in the bush; we'd race each other. He didn't like to lose."

The legendary American frontiersman Daniel Boone reportedly was able to cover forty miles a day when traveling in rough country; Hadwin would have had no trouble keeping pace. The experience of traveling overland with a fit West Coast logger would leave most people breathless and struggling. Laden with a heavy chain saw, tool belt, and cans of gas and oil, they can move through a steep mountain forest—cougar country, as it's sometimes called—with remarkable grace and speed. Some of this is due to experience and work ethic, but there is also an element of necessity: the country is so vast that unless you move quickly, you simply won't get anywhere. Often loggers will travel on catwalks they have made by felling trees end to end, enabling them to walk above the boulders and brush for a thousand yards at a stretch. Because of the rough terrain, these elevated walkways can take you thirty feet off the ground in no time, and the transition from one tree to the next is usually made by jumping, or by dancing across a slender splintered branch; they can be lethal in the rain. This is one of the reasons most West Coast forest workers wear calk boots, but Hadwin could take them or leave them. Even in winter he could be found cruising the tree line in jeans, a wool undershirt, and slip-on romeos while his colleagues would be hustling to keep up with him in a heavy parka and calks.

There is, in the life of an alpine woodsman, a heady combination of possibility and physical intensity equaled by few other occupations. For Hadwin, the mountains around Gold Bridge offered a kind of optimum challenge, a steady diet of what another B.C. timber cruiser would describe as "the unexpected heaped atop the unforeseen." Even by a forester's standards, Hadwin's work as a remote operative for the timber industry allowed him an enviable freedom; if he wanted to detour up a 9,000-foot mountain, he could, and if it presented an appealing snowfield, he could do a 4,000-foot glissade back down to the tree line. In the process, he might find a lake that had never been mapped. If he thought there might be game around, he could pack a rifle along with his compass, altimeter, and notepad. Hadwin's confidence in the woods was complete; as a result, he felt perfectly comfortable doing things that would seem suicidal to other people. Paul Bernier was with him when they ran across a pair of grizzly bears on a rockslide above Lone Goat Creek, about ten miles south of town. Instead of watching them quietly, or heading in the opposite direction, Hadwin started clapping his hands and yelling to get their attention. He succeeded, and the bears charged. Grizzlies are surprisingly fast, and once provoked, they will descend on a target with the terrifying inevitability of a furred and clawed locomotive. Lewis and Clark described encounters with these bears in which they were forced to shoot them simply to keep from getting attacked; one animal absorbed ten musket balls before it finally collapsed. Neither Hadwin nor Bernier was armed, and they had only a matter of seconds to decide what to do before the bears arrived, possibly to tear them apart. Hadwin assessed the wind direction and, with Bernier hot on his heels and the bears gaining rapidly, he dodged across a stream and feinted downwind where the shortsighted animals couldn't find them.

On another occasion, late in the fall, Hadwin set off into the mountains on a spontaneous hunting trip; despite an early snow, he was equipped with nothing but a jean jacket, a half-empty "forty-pounder" of vodka, and an open-sighted Mannlicher rifle. Two days later he returned with a mountain goat over his shoulders. Mountain goats are far more difficult to get close to than, say, deer or elk. To hit one with open sights (as opposed to a telescopic sight) is impressive under the most favorable circumstances, but even then a hit does not guarantee a kill. Hadwin, despite being not only half drunk but nearsighted, was able to track, kill, and recover the 200-pound animal alone, in winter conditions.

In addition to consuming prodigious quantities of chewing tobacco (half a tin at a time, sometimes soaked in rum), Hadwin was known for buying whiskey by the case and going on spectacular binges that, even in freezing weather, would leave him unconscious in the back of his vintage Studebaker pickup, or passed out in a snow-filled ditch, dressed only in slacks and shirtsleeves. There was a local joke: "Look, that snowbank is moving. Must be Grant." In the morning he would lurch to his feet, shake himself off, and stagger home. It is unclear why he survived (alcohol doesn't actually help a person stay warm, it merely dilates the surface blood vessels so you don't feel the chill as keenly). Early photographs show a slender, fine-boned man slightly less than six feet tall with high cheekbones and a lantern jaw; thick brown hair is parted to the side above penetrating blue eyes. Later in life he would still bear the deeply etched musculature of a man who was built for speed and distance, like a cross-country runner, or an Old World messenger.

Some who knew Hadwin during his Gold Bridge days likened his lean, sharp-eyed appearance and remote manner to James Dean's and Clint Eastwood's. There were women who admired him, usually from afar. Quiet and courteous though Hadwin generally was, he possessed an almost tangible intensity, a piercing, in-your-face conviction that some found alarming. "He always had to be the best, had to be first," his Aunt Barbara recalled. "It always had to be Grant's way. There was never any room for compromise."

And yet, compromise—of an ugly, elemental kind—lies at the root of the timber business, particularly if you are a person like Hadwin, who thrives on Nature in her rawest form. The forests he and his colleagues saw in British Columbia during the 1960s and 1970s were the same ones Alexander Mackenzie encountered nearly two hundred years earlier. They were dark, dense, apparently endless, and filled with frightening creatures; because most of the B.C. coast is inaccessible by road, it remains among the wildest regions in North America. With the exception of the occasional hunter or prospector, surveyors and timber cruisers are often the first Europeans ever to set foot in these daunting forests. Most of those who came later were only passing through because, despite its abundance of raw materials, there are very few ways to make a stable living in a place like Gold Bridge; successful mines are a rarity and most loggers were brought in from outside. But Hadwin found a way to do it: as Paul Bernier put it, he "was the de facto divisional engineer in charge of cutblock layout and design" for a vast swath of the surrounding forest. His employer was Evans Wood Products, a midsized lumber company based in Lillooet, sixty miles away. They gave him a title—Layout Superintendent—and a company truck. It was a plum of a job and one of a very few contemporary occupations that could suit a person as ferociously independent as Hadwin.

He also managed to find a woman who would put up with him; in 1978 Hadwin married a fundamentalist Christian nurse from Lillooet named Margaret, and she changed his life. He quit drinking and chewing tobacco overnight, an achievement that, given his slavish addiction to both drugs, represents an incredible feat of willpower. But what is more amazing is that he never went back. Margaret was private, retiring, and territorial; she and Grant had three children, and she was a devoted mother. The next decade would be the happiest, most stable time Hadwin had ever known. In order to house his new family, he built the most imposing structure in Gold Bridge. It was three stories tall and made entirely of hand-hewn logs; Hadwin constructed it himself from materials that had been cut, milled, and gathered locally by him or under his direction. The capstone on the oversized river rock chimney is a mattress-shaped slab of granite weighing more than four tons; the front steps, too, are a thing of massive beauty: chiseled from a single log set on an angle, the grain flows from riser to tread like a waterfall.

THE LATE SEVENTIES were boom times for the timber industry and Hadwin rode the wave; it was a good time to have a forest technician's degree, a two year ticket that he had earned in 1973. In many ways, Hadwin's mandate wasn't all that different from Mackenzie's or Lewis and Clark's: go into the wilderness, find out what is of value, and come back with a plan for extracting it. In addition to an understanding of forests and their relative commercial values, the job requires a great sensitivity to the lay of the land. Through a dense and sheer mountain wilderness, one has to be able to visualize and engineer a gentle pathway—in effect, wheelchair access—for heavy equipment. The catch, in Hadwin's case, was that if he did his job correctly, it would mean that the same wilderness he took such pleasure in exploring would soon be accessible to off-road logging trucks capable of carrying 100-ton loads (twice what's allowed on the highway) and men with orders to level the designated cutblocks as rapidly as possible. Hadwin wasn't just good at this, he had moments of true brilliance.

Wilderness road layout has become an increasingly tricky business in the past thirty years; for one, you have to "think" like a gigantic piece of heavy equipment which needs gradual inclines, strong shoulders, and a minimum of hairpin turns. But more relevant is the fact that by 1980 much of the easily accessible coastal timber was gone; left for last were the places that were hardest, and most expensive, to get to—places like Seton Ridge. "He had a sixth sense when it came to layout," recalled a field partner named Dewey Jones. "He laid out one road up this steep mountain face south of Lillooet—it was really a challenge: you'd look at the side of that mountain and say, 'There's no way you can put a road in there.' But he did; it was sort of an engineering marvel."

This is the Seton Ridge road, a twisting intestine of crushed rock and dirt that so closely follows the terrain's precipitous contours as to be all but invisible from below. A lot of timber was hauled out on it, and more than twenty years later the scars are still visible from miles away. Even under the most favorable circumstances, it takes Nature a long time to recover from a clear-cut. Known as "harvests" in the timber industry, they are shocking things to behold: traumatized landscapes of harrowed earth and blasted timber. The devastation is often so violent and so complete that if a person didn't know loggers had been through, he might wonder what sort of terrible calamity had just transpired: an earthquake? A tornado? After a few years the stumps tend to bleach out, giving the impression of headstones in a vast, neglected graveyard. Such scenes can be found throughout the Pacific Northwest, though today many of them are artfully hidden from public view by thin screens—"beauty strips"—of spared forest.

When the Hadwin family first showed up in Gold Bridge in the late 1950s, the surrounding valleys were thick with virgin, high-altitude timber. Today, as in much of British Columbia, vast clear-cuts push outward in every direction, giving the mountains the appearance of enormous animals unevenly shorn of their coats. It was Grant, in his most successful incarnation as a forest technician, who laid out many of the roads that gave loggers access to the remote forest around Gold Bridge. While doing the work he loved he helped to raze the site of many of his happiest memories. In a sense, this was a family tradition; like many older West Coast families, the Monks and the Hadwins played an active role in opening up the country Hadwin's father oversaw the massive hydroelectric dam complex that would power much of Vancouver, and his grandfather had come west to cash in on the timber boom, homesteading in West Vancouver and retiring as the proprietor of a successful logging supply company.

What is eerie is that, despite the logging industry's profound impact on our lives and on this continent, few people outside the industry have actually witnessed a logging operation. While some of the mystery can be traced to the industry's skittish attitude toward spectators, it also owes much to the average consumer's lack of interest in the origins or true costs of the resources we take for granted. Most people don't handle wood except in a finished state, and even those within the industry tend to be aware only of their particular link in the chain. If you were to ask a logger where his trees go, or a carpenter where his lumber comes from, there is a good chance that neither one would be able to tell you, and once that wood has been transformed into a chair or a paper towel, its provenance is anybody's guess. In the course of its refinement, a tree's identity devolves from a living piece of the planet, to a dead and uniform commodity bought and sold by the cubic meter, to a still more rarified product purchased by the lin-ear foot, and from there to a safe and familiar feature on our own domestic landscape that is valued less for its raw materials than for its utility and style. By then any connection to the tree it once was is as remote and abstract as a cheeseburger is to a Texas steer.

There is another reason we are so far removed from this process, though, and that is because, in most cases, the process is so far removed from us. Old-growth loggers are latter-day frontiersmen let-ting the light into the last dark corners of the country; we don't see them because they are pushing deep into places where the bulk of the population wouldn't last twenty-four hours. This is one reason the woodsman's lifestyle appealed so strongly to Hadwin, but problems arise if one stops to look down. In the timber industry, awareness causes pain. The evaluation of success involves a strange and subjec-tive calculus: at what point does the brown cloud over an industrial city become a "problem" as opposed to a sky-high banner proclaim-ing good times? When does the ratio of clear-cuts and Christmas tree farms to healthy, intact forest begin to cause aesthetic and moral dis-comfort, or real environmental damage? How does one gauge this in a place as big as British Columbia, or North America? Hadwin, like a lot of people in ethically ambiguous occupations, found his success progressively harder to live with. He was the first in his family to see the end coming, and over time he grew to believe that it had fallen to him, personally, to redress the imbalance.

CHAPTER SEVEN

The Fatal Flaw

*Midway through the journey of our life, I found
myself in a dark wood, for I had strayed
from the straight pathway to this tangled ground.*

—Dante, DIVINE COMEDY, opening lines

PAUL BERNIER DESCRIBED HADWIN as "a considerate logger and a careful road builder" who believed in taking the good with the bad, the best with the bug-ridden—when it was standard practice to just skim the cream and move on. Even his house was an effort to put his money where his mouth was. In a town—and in an industry—where people, resources, and even homes are exploited and abandoned, the Hadwin house stands alone as a kind of monument to permanence.

But as it turned out, Hadwin's timing couldn't have been worse; he was advocating restraint and moderation at a time when the logging industry was entering one of its most aggressive phases ever. The eighties were the era of the infamous Bowron Clear-cut; initiated in an effort to control an explosive pine beetle infestation, there is an ongoing debate about where containment ended and unbridled opportunism took over. In any case, the result was a starfish-shaped swath of shaved planet spreading for more than 200 square miles across B.C.'s central interior.* Local foresters described it proudly as the only man-made object besides the Great Wall of China that was visible from space. It wasn't long after this, and similar events, that B.C. was given the derogatory nickname "Brazil of the North." Since it was replanted and renamed a "New Forest," the Bowron no longer stands out quite so starkly, but it lives on as an infamous symbol of the ambiguous and codependent relationship between the provincial government and the huge multinationals that now control most of the timber industry.

By this time environmental groups had already been fighting to save big coastal trees for years, but as far as the less photogenic alpine timber around Gold Bridge was concerned, Hadwin's was a lone voice in the wilderness. "He was out of time," recalled Brian Tremblay, who has known Hadwin since they were teenagers. "He was on his own trajectory; he was talking environment and proper forest management before anybody."

One of Hadwin's duties was to prepare detailed reports on the areas where he had done reconnaissance. These documents are traditionally dry, utilitarian, and pro forma, but Hadwin began using them as a platform for critiques of logging methods and recommendations for areas he thought should be set aside. However, Evans Wood Products had hired Hadwin for his stamina and layout skills, not for his personal opinions, and his sometimes strident challenges to the status quo were not appreciated at the home office. Office politics were never Hadwin's strong suit and he could hardly be called a team player; despite their respect for his work, friction developed between him and his supervisors. His independence and isolation worked

---

*By comparison, the eruption of Mount St. Helens destroyed about 150 square miles of forest.

against him here, and as ranks closed in Lillooet, Hadwin found himself outside the circle. "I was one of the last people to see these areas
before they were logged," he would later tell a reporter. "At various
times I stuck my oar in to try to save this piece or that piece without
any success. So I guess I started to get pretty cynical."

It could be said that Hadwin's misgivings were an occupational
hazard. Timber cruisers and surveyors are avatars of the Heisenberg
uncertainty principle: woods-wise and tree-friendly as they may be,
their observations are destined to have a dramatic, if not catastrophic,
impact on the landscape. They are the last people to see the forest
intact. And yet, to try to alter this course, or even to question it within
the industry, was out of step—not just within the culture but with
the current era. Paul Harris-Jones was one of the lucky few who got
to see Vancouver Island's legendary Nimpkish Valley in all its glory.
The Nimpkish represented one of the largest stands of big timber in
the province: mile after mile of hemlock, fir, and cedar trees eight to
fifteen feet in diameter and growing as thick as cornstalks. Harris-
Jones spent an entire summer cruising the valley for Canadian Forest
Products in the early 1950s. "I was astounded by these forests," he
recalled. "It was very exciting: you'd fly into the camp on a float-
plane, take a log train to the end of the line, and then hike off into
the wilderness. The forest was so dense that our skin was paler
when we came out than when we'd gone in; for three months we
never saw the sun. The mosquitoes were god-awful; there were
floods; we fought fires. We were always trying to find a way across
the Nimpkish [River], fighting our way through this terrific jungle."

Today, the Nimpkish Valley is unrecognizable. "It was so dark and
dense and gorgeous," remembered Harris-Jones. "I came back, and it
was all gone. I couldn't believe that they had logged all but forty-four
acres—*acres!*—of the Nimpkish Valley." (Harris-Jones is now an
environmental activist and writer; in addition to finding the first
marbled murrelet nests* in British Columbia, he is credited with

---

*Like the spotted owl, this rare, semi-aquatic seabird prefers to nest in old-growth forests.

spearheading the preservation of the Caren Range old-growth forest outside of Vancouver that includes Canada's oldest known trees.)

Suzanne Simard is currently a professor in the Forestry Department at the University of British Columbia, but when she was a student she spent her summers in the mountains around Lillooet, assisting Hadwin with road layout. Her experience was similar to many others who worked with him over the years: she found him to be quiet, thoughtful, and extremely good at what he did; she was particularly struck by his almost atavistic comfort in the bush. "We'd be stumbling along, and he'd just be gone, like a coyote," she recalled. But Simard also saw what Hadwin found so upsetting. In addition to general despoliation of the landscape, landslides and the fouling of streams are among the most common side effects of mountain logging, and in an environment like coastal British Columbia, where the topsoil is thin and the rains are heavy, these problems are compounded. Evans Wood Products had a poor record in this department; in the words of one veteran forester, they were "a bottom-tier company; they gave the industry a bad reputation." In the early eighties, when Simard was assisting Hadwin, Evans took a "Bowron" approach to the forests around Gold Bridge. "It was like this big machine moved in," recalls Simard, "and began mowing it down. I can't bear to go back there now."

"We basically gutted the place," explained Al Wanderer, a second-generation logger who worked with Hadwin. "I've made a good living," he added, "but sometimes you wonder if it's all worth it."

In 1983, shortly before Evans was bought out by another company, Hadwin quit on bitter terms and went to work on his own, struggling to find a way to remain gainfully employed in the woods without "gutting" them. For three years after leaving Evans, he ran his own logging operation outside Gold Bridge where he made railroad ties by salvaging trees that had been killed by a beetle infestation that had also damaged much of the surrounding forest. "That guy worked hard," his neighbor Tom Illidge said. "It would have taken three normal men to do what he did up there." But the late eighties were a terrible time for the West Coast timber industry; the Japanese market—

crucial to B.C.—collapsed, and prices fell through the floor. Despite his superhuman efforts, Hadwin couldn't make his business pay, so he started doing freelance reconnaissance work, cruising timber and laying out roads in various places around the province. Things went fairly well until late in the summer of 1987, when, shortly before his thirty-eighth birthday, his life took a disturbing turn. Hadwin had been doing contract work for a timber company up in McBride, near the Alberta border; he had come highly recommended and Gene Runtz, the company's woodlands manager, was impressed. "He'd been doing exceptional work," remembers Runtz. "Then he left for about ten days between jobs, and when he came back it was like he was a different person—like Dr. Jekyll and Mr. Hyde. The eyes looked like they weren't there anymore. It's one of the most shocking things that's happened to me in the forest industry. He talked to us and put this religious bent on the fact that what we were doing was wrong. He said he didn't want to work for us anymore. I thought the world of the guy, but when I looked at those weird eyes peering at me—staring at me—I thought, 'Holy crappo, if this guy wants to leave—*fine.*'"

What Runtz didn't know was that while he had been away, Hadwin had had a vision. Like the monks and anchorites who once roamed from the deserts of the Middle East to the remotest outposts of the British Isles, Hadwin had ventured into the wilderness and received a message that he could not ignore. As the theologian Benedicta Ward wrote, "The essence of the spirituality of the desert is that it was not taught but caught." Hadwin was not searching for such an experience, it came up from behind and clubbed him over the head. The episode passed as mysteriously as it had come and Hadwin went back to freelancing, where he continued to receive excellent work reviews, but his supervisors could see there was something different about him. "The amount of work he could do on his own was incredible, and his plans would be great, but at times you'd think he was sort of obsessed," recalled Grant Clark, who supervised Hadwin a year later outside of Kamloops, a three-hour drive east of Gold Bridge. "He would stay out there; he wouldn't come back to town. He was always at arm's length: you could say he was doing great work, but it didn't mean anything to him." To Clark, it seemed as if Hadwin was operating at a different level. "He seemed to be in tune with actual nature; he always knew exactly where he was. Animals would stay close by; he wouldn't spook them."

But as competent and in tune as Hadwin may have been, the implications of the incident at McBride were ominous; it appeared that after a twenty-year hiatus, the family ghost that had killed his brother had set its sights on him. It was hard to imagine the impervious Grant being vulnerable to anything his brother had been because, in every other way, the two couldn't have been more different. Where Grant was always lean and wiry, Donald, who was twelve years older, was almost pretty, with full red lips, round cheeks, and wavy blond hair. He had been an altar boy while Grant had been a hellion. Grant's first day of kindergarten ended early when he was sent home in a cab with a note pinned to his sweater that said, "Do not send this boy back." "He was like twelve kids," recalled a cousin, "and smart as a whip." But he would never be the white-collar professional his father had hoped for. Donald, on the other hand, appeared to make the grade. He was more submissive, toeing the same line that Grant would continually push against, and with his father's stern encouragement, he tried to follow Tom Hadwin's tough act, entering the University of British Columbia's electrical engineering program. He did well enough, but the satisfaction was short-lived; Donald left home as soon as he could, seldom returning for visits.

Then, a year before Grant went logging, Donald resurfaced. He had no friends and no job, only a diagnosis: paranoid schizophrenic. Despite his family's best efforts, he refused treatment of any kind. There is little doubt that this horrific undoing of his formerly successful brother helped to drive Grant away from the professional mainstream and into the woods. Indeed, subsequent events may have made the forest seem like a far safer—and saner—place to be. In February of 1971, the same year Hadwin went back to school for his

forest technician's degree, Donald had a final dinner with his parents
in their West Vancouver home. Afterward, he headed back downtown
across the Lions Gate Bridge, Vancouver's answer to San Francisco's
Golden Gate. Halfway across the span, practically within sight of his
parents' house, Donald paused. Surrounded by hulking mountains
and shimmering water, he climbed over the railing and jumped. He
was thirty-four years old.

When Hadwin quit his job at Evans, he was in his mid-thirties,
too. He was opinionated and eccentric, but he was also a strenuous
provider, a helpful neighbor, and "a hell of a nice guy." He was the
kind of father who remembered his children's birthdays when he was
away; when he was home, he would take his kids fishing and snow-
mobiling and help them with their math homework. As Tom Illidge,
put it, "He wasn't lazy and he wasn't crazy." Illidge is one of Gold
Bridge's oldest and most successful residents, one of the few who
stayed put, stayed sober, and prospered there. He sympathized with
Hadwin's growing disdain for the company men who wield so much
power over the forest without knowing their way around in it: "Half
of those assholes have never been four feet from a parking meter in
their lives," he said

But while Illidge, Al Wanderer, and Hadwin's other colleagues
were able to swallow their irritation and press on, Hadwin eventually
found it intolerable. Late in 1989 his mill was vandalized and this
exacerbated an emerging tendency toward paranoia. Sensing that his
neighbors were turning against him, he moved his family out of
Gold Bridge; shortly afterward, his freelancing contracts dried up.
The Hadwin family relocated to Kamloops in the high, dry ranch
country of south central B.C. With 80,000 people Kamloops was
barely a city, but compared to Gold Bridge, it was a teeming metrop-
olis. While it had better schools for the children and more job opportunities, it was a terrible place for Grant; as his former assistant,
Suzanne Simard, put it, "Moving [him] to Kamloops would be like
taking a bear and putting it in a zoo." Out of his element, Hadwin

struggled to find meaningful work, sending out letters and résumés, and advocating on behalf of friends and neighbors he thought were being ill used by the system. Unemployed except for sporadic volunteer work at a local retirement home, he had a lot of time on his hands, and he began writing letters on a wide range of issues to political figures all over Canada and the world. In a letter to a provincial supreme court judge he wrote:

> The Forest industry in british columbia, appears to be one example, of economic remote controlled TERRORISM, on this planet, with professionals leading the way, in "severe symptoms of denial, that there is any problem."

Later, in a widely distributed two-page memo entitled "A Few Thoughts About University Trained Professionals and Their Equivalents," Hadwin enumerated his observations about the professional class, including the following:

> 3. Professionals appear to "DENY" or ignore "The Negative," particularly about themselves or their projects.

> 4. Professionals appear to create and positively reinforce facades and perceptions until these facades and perceptions are "perceived" to be fact (media do this all the time).

> 7. "NORMAL" today appears to be "professional values" rather than say "Spiritual Values" or a reverence for life.

In 1991 Grant and Margaret separated, and she got custody of the children. In early 1993, increasingly frustrated and unable to handle the pressures he felt in Kamloops, Hadwin headed north on a rambling hegira through the Yukon and Alaska, where in early June he sought refuge on a remote island. A month later Hadwin was stopped at the United States border with three thousand hypodermic needles in the trunk of his car. He talked his way through customs and proceeded to Washington, D.C.; once there, he presented himself as an advocate of needle exchange and safe sex, distributing needles and condoms to anyone who wanted them; he also donated thousands of dollars to a local food bank and homeless shelter. In July, with two thousand needles remaining, he proceeded to Miami, where he caught a plane to Moscow; from there he continued eastward, donating needles to children's hospitals as he went. He was arrested by the police in Irkutsk, Siberia, but apparently finessed the interview and parted on good terms. Hadwin, however, wasn't simply on a goodwill mission, he was also looking for work; Siberia is one of the few places in the Northern Hemisphere whose forests rival British Columbia's.

When Hadwin returned to Kamloops, people who knew him were alarmed by what they saw. The guerrilla theater dress he sported on his travels (running shorts, riding crop, boots with spurs, and a baseball cap festooned with needles and condoms) raised some questions about his mental state. His apparently stress-induced paranoia had also begun to blur with reality as he found himself in situations in which people really were out to get him. That October, on the same day he was served with papers limiting his visitation rights to his children, Hadwin got into a running altercation with the driver of a semi on the Trans-Canada Highway. It degenerated into an almost comic scene with the mammoth Peterbilt tractor chasing after Hadwin in his little Honda Civic. The truck driver refused to give up and rode Hadwin's bumper all the way back to his wife's home, at which point both men jumped out of their vehicles and got into a violent argument. The truck driver was four inches taller and fifty pounds heavier than Hadwin; his hands were balled into fists and a fight seemed imminent. Hadwin ran up the driveway, snatched up a two-by-four, and shouted at the driver to "get the fuck out of here!" Then he clubbed him once in the head. The man collapsed, at which point Hadwin proceeded, immediately, to help him back to his feet. When the truck driver and his wife waved him away, Hadwin drove to the police station and turned himself in. It was his first-ever brush with the law.

Hadwin was sent to a forensic hospital for a monthlong psychiatric evaluation where he was interviewed extensively by several doctors. Although all of them found evidence of what one psychiatrist termed "paranoid reaction," the only diagnosis they could agree on was that he was mentally competent and fit to stand trial. He was given a prescription for a low dose of antipsychotic medication and his condition improved dramatically. It is hard to say whether it was the medication or Hadwin's internal cycle that was responsible for the improvement because it is unknown how often he took the medication, if at all. Within a couple of months he had secured a job at a local lumber mill, working as a veneer peeler (for plywood), and he turned in a twenty-page report for a proposed logging road. Hadwin worked on the project alone, and when his boss, Pat McAfee, asked him if he had an emergency contact person in case of an accident, Hadwin replied, "If I can't get out of the bush on my own, I don't want to come out."

"He was very proud of his work," recalled McAfee. "He was one of the best layout contractors I've seen."

That September, just short of his forty-fifth birthday, Hadwin placed second in a thirty-mile cross-country "ultra-run." When his trial date came up, he was offered a plea bargain; he pled guilty to assault and was placed on a year's probation, during which he regained custody of his two oldest children. He took the two teenagers to visit Cathedral Grove on Vancouver Island, where they posed for pictures in front of giant cedar trees. It was during this period that Grant Clark, Hadwin's former supervisor, had a harrowing encounter. "I saw him downtown around 1995," Clark recalled. "The eyes looked hollow, like they were looking through you. He didn't know who I was. It was eerie to see a guy with *so* much bloody talent sunk to those depths."

By now Hadwin had been riding a neurochemical roller coaster for at least seven years. In spite of this, he managed to honor the terms of his probation, but being kept on such a short leash put enormous strain on him. Like a lot of men who feel their freedom and purpose have been somehow denied, Hadwin began casting about for other ways to manifest his competence and make an impact. He stayed up-to-date on local and international news and became increasingly involved in environmental and native issues, both of which are highly contentious topics in British Columbia. During an armed standoff between Indian activists and the Royal Canadian Mounted Police (RCMP) that took place during the summer of 1995 at Gustafsen Lake, 100 miles north of Kamloops, Hadwin went so far as to travel there and offer his services as an intermediary. Not surprisingly, his offer was declined. The Gustafsen Lake standoff, during which two RCMP officers were shot in the back, was one of a number of such skirmishes occurring throughout B.C. and the rest of Canada at this time, and it received national news coverage. The impasse, which ended in a surrender after three weeks, made a deep impression on Hadwin, and he sent dozens of faxes and certified letters to native groups, politicians, and the media. To CNN, he wrote:

> Your focus appears to be Bosnia and O. J. Simpson. Your Native American problem, however, parallels our own and yet your coverage, appears to be nonexistent. . . . You would apparently go to any lengths to deflect the focus from the real issues, which discredit yourselves or your professional institutions.

Hadwin's letter-writing campaign continued to intensify, and within this raft of correspondence is what appears to be a final attempt to find meaningful employment. On January 12, 1996, in response to an ad for a Forest Renewal Project coordinator, Hadwin sent his still-formidable résumé, along with the following cover letter:

> I do not like clearcutting and my philosophical differences,

with the Forest Industry, run deep. If you are prepared to try a "gentler approach," to forestry, with less "short term profit," I may be able to help. I am not familiar with the new "buzz-words," such as Forest Renewal. All of Forestry and most of the Forests, appear to need "Renewing," in some form or another.

Hadwin didn't get the job. His only consolation, it seemed, would be a woman named Cora Gray. One of Hadwin's neighbors in the apartment complex where he was living was Matilda Wale, an elder from the Gitxsan tribe whose homeland borders Tsimshian territory, due east of the Queen Charlottes. The Gitxsan are inlanders; as a result, they remained insulated from Europeans for much longer than the coastal tribes. Even as recently as the 1920s, it was considered unsafe for government officials to travel there. Hadwin looked after "Tilly" Wale, helping her out when she needed it and occasionally buying her groceries. In July of 1996, Wale's half sister, Cora Gray, came through town on her way to a powwow, and Hadwin put her up in his apartment. Gray was in her mid-seventies and she had lost her husband and her mother within the year; she was lonely and had a kindly, forgiving manner. She reminded Hadwin of a favorite aunt and he took a liking to her right away. Gray had a camper van and in it the two traveled as far as the Salmon Glacier, an enormous ice-blue tongue that laps the headwaters of Portland Canal, a hundred-mile-long fjord on the B.C.-Alaskan border. Considering their radically different backgrounds, they had a lot in common; both had been separated from their families at an early age and sent away to schools they hated—Gray to a residential Indian school in Alberta, and Hadwin to a British-style boarding school in Vancouver; physical abuse was commonplace in both institutions. The wounds left by these early banishments gave each an understanding of the other that was rare to find in such an incongruous match. On their trips together, Hadwin would often take time out to go running and swimming; then, in the evenings, he would cook for the two of them. They played cribbage and rummy, and laughed a lot; before long Gray had become Hadwin's closest friend and confessor. Gray was with Hadwin when he made his first trip to the Queen Charlotte Islands.

THE QUEEN CHARLOTTES, all but forgotten since the collapse of the otter trade, had been rediscovered during World War I. This time it wasn't furs, fish, or gold the outsiders had come for, but airplanes. In places like Washington's Olympic Peninsula, Vancouver Island's Clayoquot Sound, and in the Yakoun Valley, they literally grew on trees, specifically, big old Sitka spruce. Prior to the war, Sitka spruce was a low-value tree, frequently passed over in favor of two other Northwest species: Douglas fir (also known as the "money tree"), which had become the builder's choice for framing, flooring, and trim, and western red cedar, whose water-resistant properties were much sought after for shingles, siding, and fence posts. The only reason for cutting a Sitka spruce down was because it stood in the way of a cedar; once on the ground, it would often be left to rot or, if convenient, maybe pulped for paper.

But when early airplane designers discovered the tree, all that changed; the lowly—but huge—Sitka spruce became an aristocrat overnight. Light in weight, Sitka spruce wood possesses a rare combination of strength and flexibility that is ideal for making airplane wings and fuselages; cut into strips and laminated, it also makes excellent propellers. It has an added benefit in that it doesn't splinter when hit by bullets—an unusual quality for any harder wood. For these reasons, the highest grade of Sitka spruce became known as "airplane spruce," and the Charlottes had one of the highest densities of it anywhere on the coast. During the war years these trees were so sought after that they became the object of an extraordinary mobilization of military forces. Starting in 1917, more than 30,000 American soldiers from the hastily formed Spruce Production Division, along with thousands of Canadian loggers contracted by Britain's Imperial Munitions Board, were sent into the coastal forests to cut and mill trees for the war effort. Much of the wood harvested by these "spruce soldiers" went to build French, English, and Italian warplanes.* By the time the Germans surrendered, less than two years later, enough spruce had been harvested to girdle the earth one and a half times (about 200 million board feet). However, buried in the Commission on Conservation's *Tenth Annual Report* from 1919 is a sobering glimpse of the future of West Coast logging:

> The supply of Sitka spruce suitable for aeroplane construction
> is extremely limited. . . . [and the] continuance of cutting on a
> war basis for another year would have practically exhausted
> the spruce which should be secured at a reasonable expense of
> money and effort. . . . Only the large trees contain the clear,
> fine-grained lumber required, and these cannot be replaced in
> centuries. Most of the aeroplane material was cut from trees
> 500 to 800 years old, and it is doubtful if the succeeding stands
> will ever attain the same quality as these virgin stands.

While concerns like this were raised periodically over the ensuing decades, it would be more than fifty years before any meaningful action was taken. By then many of the islands and much of the coast would be reduced to moonscapes.

The spruce soldiers' highly organized assault on the coastal forests helped to usher in the modern age of logging when the technology for dismantling forests began outstripping the imaginations of those who wielded it. It also led to phenomenal waste: with less desirable

---

*Twenty years later, during World War II, the British-built DH-38 Mosquito, made almost entirely of Sitka spruce, Douglas fir, birch, ash, and Ecuadorian balsa, was the fastest, most versatile airplane in the Allies' arsenal. Variously armed with reconnaissance equipment, cannon, or machine guns, it could also carry a 4,000-pound "blockbuster" bomb. Not only did it suffer the lowest loss rate of any Allied warplane, it was also the easiest and cheapest to repair. The lightweight fighter-bomber was so fast that the Americans issued standing orders for their swiftest plane, the P-38 Lightning, never to be flown alongside it. Despite being powered by propellers, the Mosquito had a top speed of more than 400 mph (unloaded), making interception all but impossible by any other aircraft. Its headline-friendly name notwithstanding, Howard Hughes's famous *Spruce Goose*, the largest airplane ever built, included only a small proportion of Sitka spruce.

A battle between the American fur trading vessell, *Columbia*,
and Kwakiutl warriors in Queen Charlotte Sound, 1792

Nuxalk canoes from the central B.C. coast, 1914 (from a dramatization by
photographer Edward S. Curtis).

The full outfit of a north coast warrior.

Haida war dagger with an eagle crest on the pommel. Collected at Masset by A. Mackenzie, 1884.

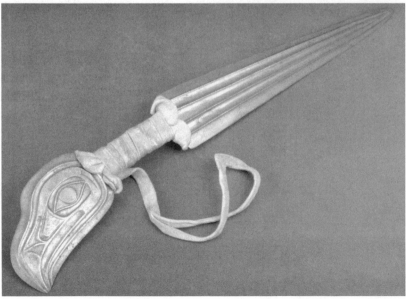

This dance mask represents a *gagiid*, a person trapped between the surface and spirit worlds by the experience of nearly drowning at sea in winter. Collected on Haida Gwaii by Israel W. Powell, 1879.

Ox team hauling logs on a "skid road" through what
is now downtown Vancouver c. 1900.

Loggers on a "railroad show" with a freshly loaded fifty-ton log, c. 1935.

Haida fallers c. 1925, posing on a springboard with one of the
thousands of huge Sitka spruce that once stood on Haida Gwaii.

A high rigger topping a spar tree (a) and taking five (b), c. 1925.

Faller watching for widowmakers in winter rain.

A feller buncher represents the future of logging:
smaller trees and bigger machines.

Clearcut with logging road and subsequent erosion, Haida Gwaii.

Grant Hadwin returning from a swim
at Rushbrook Floats, Prince Rupert, January 6, 1997.

Grant Hadwin embarking from Prince Rupert
by kayak, February 12, 1997.

Border crossing, Hyder, Alaska, a place Grant Hadwin visited several times.

Leo Gagnon, Tsiij git'anee chief-in-waiting (right),
and his son, with the stump of the Golden Spruce, October 2001.

Mortuary poles at Nan Sdins, UNESCO site,
Gwaii Haanas National Park Reserve, Haida Gwaii.

Memorial pole for Ernie Collison (Skilay), spokesman for the
Golden Spruce. The pole was carved by Chief Jim Hart and
assistants, and erected by the community on March 29, 2003.

species such as hemlock and balsam being abandoned in favor of their more profitable neighbors, it has been estimated that, on average, nearly 30 percent of a cutblock's usable wood was left to rot—or burn—among the slash. Despite having grown up in the forest, many West Coast loggers seem surprised at how fast their trees have fallen, and some of this can be attributed to a kind of magical thinking that was at work in the woods. Many in the industry were operating under the untested assumption that by the time the old growth was gone, the next generation of trees would be ready to harvest. This might have been true in the hand-logging—or even the steam-logging—days, but no longer. By now the industry had become blazingly efficient, and yet the majority of cutover forests were still left to reseed themselves once the loggers had gone.

Following a local catastrophe like a fire, windstorm, or clear-cut, a forest will rebuild itself through a natural process called "succession"; on the coast, a series of species, beginning with berry bushes and low scrub and progressing through fast-growing alders to shade tolerant "climax" species like spruce, cedar, and hemlock, will follow one another in a predictable pattern that can take centuries to unfold. In the Charlottes, the seemingly commonsense practice of actively replanting cutover areas was not institutionalized until the 1960s; for interior and mountain forests like those around Gold Bridge, replanting wasn't introduced until the 1980s.

"They bullshitted us," a retired Haida logger named Wesley Pearson said of the logging companies he used to work for back in the fifties, sixties, and seventies. "They said when we finished logging [the Queen Charlottes] we could start over. Well, we logged it a hell of lot faster than anybody thought we'd log it. A lot of mistakes were made; the government didn't keep an eye on the big companies."

But lies are easier to swallow when the money is good, and for fallers and high riggers like Pearson, it was. Another Haida faller named Bill Stevens could have been speaking for the entire logging industry when he said, "When you have that job, you forget about everything else for awhile."

To get an idea of the scale of logging taking place in the Charlottes during the last thirty years, one need only look as far as the *Haida Monarch* and the *Haida Brave*. At the time of their launching, they were the world's largest floating log carriers, and both were built to serve the islands; the *Monarch* (the larger of the two) is capable of carrying nearly four million board feet of timber (about 400 truck-loads) at a time. When one of these vessels dumps its load at the booming grounds in Vancouver, it can generate a spontaneous wave eight feet high.

MacMillan Bloedel, Canada's biggest logging company, operated the *Haida Monarch* and *Haida Brave*, and both vessels would routinely travel through Masset Sound on their way to Juskatla Inlet and the "log sort" where much of the Yakoun Valley's trees were taken for transport. Whereas the nineteenth-century otter ships had appeared to be in plausible proportion to the houses and canoes that lined the village beaches, these modern vessels dwarf everything in sight. Both ships are approximately four hundred feet long, and they seem a tight fit in the slender gut of Masset Sound. From beach- or boat-level, a log carrier looks less like a ship than a floating wall of steel; more than a hundred feet above the water loom fifty-ton cranes for loading the literal forests of cargo. In addition to their tactless choice of names, their color scheme exudes a sinister otherworldliness: the vessels are matte black, like Stealth bombers, as if they were designed to evade detection. But the opposite effect is achieved: as they emerge from the fog, moving inexorably across the celadon-gray North Pacific, they bring their concentrated darkness with them. Watching them approach, one has the feeling that no good can possibly come of this. And yet, laden with a suburb's worth of cedar decking, these seagoing resource removers are as much a part of the modern Haida legacy as the Nor'westmen. For decades, sometimes as much as twice a week, the residents of Masset have watched their patrimony get loaded up and carried off. Although 80 percent of the logging jobs are given to people from off-island and 95 percent of the wood is sent south, some of them have profited from it. As Wesley Pearson put it,

"If you're born in the Charlottes, you're either a fisherman or a logger." These are, after all, virtually the only jobs available out here, and so, many Haida find themselves in a strangely familiar double bind: aid and abet the plundering of their historic homeland, or get left behind.

"I liked logging as a young man," recalled Pearson. "I'd never have made anywhere near the money I made in logging 'cause I didn't have any schooling. I can't say anything against it 'cause too many people depend on it. But how do you control it? The big companies always get the wood they want."

This question had been bothering Hadwin, too, and he carried it with him everywhere he went. In September 1996, shortly after what came to be known as the *Haida Brave* blockade, Hadwin visited Haida Gwaii for the first time. In so doing, he stumbled into a vortex of conflicting hopes, dreams, and ambitions. The golden spruce and its environs embody the collision between the ideal of "Haida Gwaii": a true rainforest paradise complete with gigantic trees and genuine Indians, and its industrial alter ego, "the Charlottes," an offshore timber warehouse employing vertical storage. British Columbia has been described as a banana republic, only with bigger bananas, and nowhere else in the province is this more blatantly the case than here. As such, the islands are an internationally recognized poster child for loggers, environmentalists, and native rights activists alike. The golden spruce was caught in the middle.

A month earlier, on August 1, while staging a protest against Mac-Millan Bloedel's continued logging of the coastal rainforest, Greenpeace activists had been fire-hosed off the deck of the *Haida Brave* while it lay at the dock in Juskatla Inlet. Later that afternoon, as it made its way down Masset Sound with a full load of logs, the barge was intercepted and forced to turn around by approximately fifty Haida in war canoes and motorboats. This wasn't the first time: with the assistance of environmental groups, the Haida had staged several highly successful anti-logging campaigns during the 1970s and 1980s; not only did they establish the archipelago as a key battleground in the coastal forest wars, but they marked it as the site of one of the earliest—and greatest—victories for forest preservationists. The creation of the Gwaii Haanas National Park Reserve in 1987 saved the southern third of the archipelago from logging but also resulted in the loss of more than a hundred forestry jobs. What distinguished this latest event was that it moved the conflict "offshore"; it was the first time in nearly a hundred and fifty years that the Haida had taken action against a foreign vessel.

With the point having been made, the *Haida Brave* was allowed to pass the following day, but Greenpeace continued to harass the barge as it made its way south, and just outside Vancouver some activists successfully boarded, chaining themselves to logs and loading cranes. The islands are a small place and the Haida and Greenpeace actions were big news there. For six weeks afterward, the Greenpeace boarding and its legal aftermath received steady coverage in B.C. newspapers. It is highly likely that Hadwin would have been following this story, just as he had followed the standoff at Gustafsen Lake.

Hadwin had been at home in Kamloops when he made the decision to go to the islands, but because Hazelton, Cora Gray's hometown, was on the way, he took Matilda Wale and her boyfriend with him for a visit. Once in Hazelton, Hadwin invited Gray, along with her sister Martha and her husband, to join them in the islands, and he paid their way across. They would have made an odd party: five elderly Indians and Grant, still rippling with muscle, still casting about for a task that was consistent with his principles and equal to his stamina. Hadwin and his entourage spent a week in the islands, and during this time he visited the golden spruce. The tree's location couldn't have been less auspicious: over the years it had been surrounded by a maze of logging roads, and all of them terminated at Juskatla, the base of modern logging operations in the Queen Charlottes. Lying only a few miles west of the golden spruce, Juskatla is a chilly, desolate parallel universe of scoured earth, heavy machinery, and dozens of white Ford 250 pickups—the official vehicle of the logging industry. Equipment is stored and repaired there in a cavernous building, and off-road logging trucks rumble in and out like clockwork. On one typical day, a large sign proclaimed 9 DAYS WITHOUT AN INJURY. Nearby is the dock and log sort where the barges are loaded. Evans Wood Products had a yard like this on the outskirts of Lillooet, and once you've seen one, it's not hard to understand why a person like Hadwin would avoid it at all costs.

The golden spruce grew equidistant between Juskatla and Port Clements, the pioneer settlement that has since evolved into a bedroom community for loggers; today, about 530 people live there. Located halfway up Graham Island, on the eastern shore of Masset Inlet, the village's welcome sign is made from an uprooted cedar stump, and once you're past it, the first thing to greet visitors is a clear-cut littered with slash piles and rusting logging equipment. It is so wet here that any object capable of casting a shadow is also a breeding ground for algae; if left alone, the green slime will give way to mosses and ferns; eventually seedlings will take root and, in less time than just about anywhere else outside the tropics, an abandoned truck, or a mobile home roof, will become an ecosystem unto itself. It could be argued that the golden spruce owes its preservation to the village's loggers and foresters. Over the years, locals had grown fond of the curious tree; Harry Tingley had picnicked by it with his father in the 1930s, and it had always been a place where islanders would take friends and family who were visiting from the mainland. For Haida and Anglo alike, the tree was like an old friend, a benign and reassuring constant for all who knew it.

To this day, come October, Tsiij git'ance clan members gather on the Yakoun River, downstream from the golden spruce, in order to catch salmon as they make their annual trip up to Yakoun Lake in order to spawn and die. There is every reason to suppose that this seasonal harvest has been performed in roughly the same place, using roughly the same technique, for millennia. It's almost dizzying to imagine the dozens—perhaps hundreds—of generations who have participated in this unbroken cycle of food gathering. Today the entire Tsiij git'anee clan could fit inside a two-car garage, but at one time the clan controlled a significant portion of the Yakoun River watershed, including the spot where the golden spruce stood.

Before European settlers and miners arrived in the 1860s, the Queen Charlotte Islands were exclusively Haida territory, and fishing, hunting, berry picking, or water rights to a particular area were held by one clan or another. For this reason among others, intratribal warfare and territorial disputes were a fact of life in the archipelago, just as they were in the Hawaiian Islands. Thus, the land the Tsiij git'anee claim now may not always have been theirs and its ownership is still in question. Their claim has been contested by a Tsiij git'anee offshoot called the Masset Inlet Eagle clan, but this claim has had to take a number behind several other claims: that of the Haida Nation, the Canadian government ("the Crown"), and MacMillan Bloedel, Ltd.

Until recently, M&B had major holdings in Europe, Southeast Asia, South America, and the United States. Its Canadian properties included an enormous lease in British Columbia called Tree Farm License (TFL) 39 which is composed of timberlands on the mainland, Vancouver Island, and the Queen Charlottes. MacMillan died in 1976, and in 1999 the company was taken over by Weyerhaeuser, the world's largest wood products company, based in Tacoma, Washington. Weyerhaeuser, which has been a dominent force in the timber industry for well over a century, controls timberlands throughout the world. The Charlottes' portion of TFL 39 is called Block 6, or the Haida Tree Farm License; it encompasses many of the archipelago's northern islands and much of the Yakoun Valley, including the golden spruce.

By the time Hadwin showed up, most of Block 6 had been leveled at some point during the previous eighty years; entire islands had been shaved bald, in some cases, out of spite born of intraisland rivalries. Much of the landscape has been permanently scarred by the landslides that follow poorly managed clear-cut operations. It is hard to appreciate the scale of the logging until you see it from the air. "When you fly over the northern islands now and see all that's been taken," said a Haida artist named Hazel Simeon, "you can't speak for a few days afterward." As a result of all this activity, the golden spruce was one of the few mature Sitka spruce trees still standing at the north end of the Yakoun River, and as such it had become even more of an anomaly than it already was. Most of the other survivors, including some big cedar and hemlock, were clustered around it; together, they composed a tiny island of old-growth forest in what is, effectively, a huge clear-cut in various stages of recovery.

In the late 1960s, MacMillan Bloedel began reserving small patches of forest that were considered particularly beautiful or environmentally sensitive. These "set-asides" were generally minuscule, seldom amounting to more than five or ten acres—nowhere near big enough to serve a significant conservation function for the ecosystem. Their primary purposes were recreational and symbolic—the briefest of nods to the great forests that had once stood there. One drawback to these pocket parks is that with no other tall trees to protect them, they are extremely vulnerable to "windthrow" (getting blown down). Even today, these little reserves are begrudged by some in the industry; in 2003, during a stroll through an old-growth set-aside on Vancouver Island, a local forester confided, "If this was mine, I'd cut it all down and plant it in fir." The park covers eight acres. A hundred yards away, loaded logging trucks were passing at a rate of one every twenty minutes.

Recreational possibilities were seen for the golden spruce in the mid-1970s and at the urging of the local logging and forestry community, M&B planned to set aside twelve acres of old-growth forest around the tree. However, the tree's protection became a moot point when recent environmental regulations declared riverbanks and other sensitive riparian zones off-limits to logging. The Haida were not formally consulted on the matter because, apparently, there was no awareness among the white community that the tree had any special meaning for them. But the same could be said for the Haida themselves; even within the tribe, only a handful of people knew the story associated with it, and at that time they had more pressing concerns. Indians did not get the right to vote in Canada until 1960, and the Haida's resurgence was still in its infancy. For them as for many other North American tribes, a period of dramatic rediscovery was just beginning.

Meanwhile, MacMillan Bloedel ran a proper trail in to the golden spruce, and a bench was built there so that visitors could view the tree, which stood across the river on the west bank. The tree itself was inaccessible unless one had a boat, or walked a mile up to the nearest bridge and then back downriver on the other side, a detour that takes several hours due to heavy brush and windfalls. In 1984 tour buses began making regular stops at the tree, benefiting local businesses, including the Golden Spruce Motel. In 1997 the town's growing ecotourist trade got an additional boost when an albino raven showed up. Usually killed or ostracized by their black counterparts, the white raven was the only one of its kind in the province. Between it and the golden spruce, Port Clements had cornered the freak-of-nature market in western Canada.

Both creatures had a startling, supernatural quality to them, and on a sunny day the golden spruce's luminescence never failed to impress—or to mystify. D'Arcy Davis-Case, a forestry expert who lived in the Queen Charlottes for years before becoming a consultant to the United Nations on forestry issues, recalled that "botanists and dendrologists were always trying to explain the tree's golden color." When asked what they had concluded, Davis-Case smiled and rolled her eyes. "Magic!" she said.

To those who were lucky enough to see the golden spruce in bright sunshine, Davis-Case's explanation sounds plausible enough. Many spoke of its peculiar radiance, as if it were actually generating light from deep within its branches. Ruth Jones, a Vancouver-based artist, visited the golden spruce late one sunny afternoon in 1994. "It looked as if it were made of glowing gold," she said. "It was like a fairy tale: how can this be?" After seeing it one day in 1995, a journalist named Ben Parfitt came away feeling that it was "somehow closer and more alive than all of the other trees around it." Marilyn

Baldwin, the owner of a sporting goods store across Hecate Strait, in Prince Rupert, visited the tree on a gray foggy day in the early 1990s. "A few minutes after we got there," she recalled, "the sun burned the fog off, and suddenly there it was in its golden brilliance. We called it the Ooh-Aah tree, because that's what it made us all say." A senior engineer for M&B who saw the tree under similar circumstances compared its sudden illumination to a religious experience.

But Hadwin saw something different, and it was the same thing that many of his more pragmatic colleagues saw: a "sick tree." More so than most people, he would have been struck by the contrast between the vestigial grove containing the town mascot and the free-range saw log farm that surrounded it. To a person who knew the woods as well as Hadwin, it would have been as jarring and ludicrous as an albino buffalo on a putting green. Where were all its healthy counterparts? Headed south on the *Haida Brave*.

CHAPTER EIGHT

The Fall

A fool sees not the same tree that a wise man sees.

—William Blake, PROVERBS OF HELL

NOT LONG AFTER his trip to the Queen Charlottes with Cora Gray, Hadwin left Kamloops for good. He headed north again, alone, eventually winding up at the Yukon Inn in Whitehorse. Whitehorse, the territory's capital, is situated just north of the B.C. border where the Yukon River drops down from its source near the Chilkoot Pass to begin its 2,000-mile arc up through the vast heart of Alaska and down again into the Bering Sea. Winter lasts a long time here and it had already begun by the time Hadwin arrived. He had always taken pride in his high tolerance for frigid water, and over the years he had swum in icy rivers throughout B.C., Alaska, the Yukon, and Russia. While he was in Whitehorse, he began taking dips in the Yukon River. The banks were covered in snow by this time, and ice was starting to form on the surface. Besides swimming and exercising, Hadwin's purpose for being in Whitehorse was unclear, but he stayed in regular phone contact with Cora Gray. He missed her company, and in mid-November he persuaded her to join him, going so far as to pay for her plane ticket.

Gray was watching as Hadwin went into the Yukon one December day when the air temperature was thirty below zero. By this time the only part of the river that wasn't frozen was a particularly fast-moving section by a dam outflow. Hadwin walked out onto the ice and used a stepladder to lower himself in; he remained immersed for about fifteen minutes. Witnesses were so alarmed that the Mounties were called, and a reporter from the *Yukon News* showed up as well. "The water was smoking," Gray recalled. "When he got out, there were icicles hanging off his eyebrows and hair. He ran back to the car, where I was waiting, and he said, 'I know I'm OK when you're there watching me.' I asked him, 'Why are you torturing yourself?' and he said, 'I'm training myself. I won't be around here next year.' I knew he was planning something."

But she had no idea what. During the previous six months Hadwin had confided in her the most intimate and painful details of his life, but his future plans were a mystery. Gray began to get nervous; she had intended to stay in Whitehorse for only two weeks, but under pressure from Hadwin, she ended up staying for six. On several occasions local Indians took Gray aside and told her they had a bad feeling about Hadwin, that she should get away from him. "When I mentioned flying home," Gray explained, "Grant cried like a baby, saying, 'I think you're the only one who's ever worried about me.'" But he also told her not to answer the phone when her sisters called. "Finally, I persuaded him that I had to go home, and he offered to drive me. He said, 'Don't tell your sisters you're coming home; surprise them.'"

It was at this point that Gray began fearing for her life. This was also when Matilda Wale had the frog dream. Like the Haida, the Gitxsan are divided into clans; Gray is a member of the Frog clan, and so is her half sister, Matilda Wale. Only days before Gray was due to leave Whitehorse, Wale dreamed of a frog getting crushed by a car. She was so frightened that she called Gray and told her. Gray was frightened, too, but she was so far from home; there was nothing she could do.

They left Whitehorse at 4 A.M. on December 30. It was a fifteen-hour drive through extremely remote country to Gray's home in Hazelton, but due to the high latitude and the time of year, the sun would be up for only six hours of the trip. Moose, wolf, cougar, and bobcat are common sights up here, and their disembodied eyes glinted green and orange as they stared back from the darkened road-side. At five-thirty that afternoon, two hours north of Hazelton, they reached the Nass River Bridge. Like most bridges in the north, this one is only one lane wide and Hadwin headed toward it at full speed. Despite the bright moonlight and clear sight lines, he failed to regis-ter that a pickup truck was crossing from the other side. Gray remem-bers being very calm, saying, "Grant, did you know it's a one-lane bridge?" The road was icy at the entrance ramp, and at the last minute, Hadwin hit the brakes. His Honda skidded and went side-ways, up onto the low railing. From the passenger seat, Gray contin-ued narrating what she believed was the end of her life: "Then I said, 'We're going into the Nass.' I didn't panic; I just thought, 'If I'm going to go, I'm going to go.' I was kind of expecting it."

In the end, they didn't go into the Nass; they hit the pickup head-on. Gray's ankles were shattered, her cheek was broken, and both hands were bruised through; Hadwin, meanwhile, suffered only a cut lip. Even more worrisome than Gray's injuries, however, was the fact that it was forty below zero and they had lost their heat source. At this temperature cast iron can shatter like glass, exposed flesh will freeze in moments, and the touch of metal will burn like fire. The nearest ambulance was two hours away. Hadwin jumped out of the car to assist Gray, but in his haste he forgot to put on his gloves, and when he attempted to open her door his fingers burned instantly. Meanwhile, the suitcases in back had been thrown against the pas-senger seat and Gray was indeed being crushed—she was choking on her seat belt—but Hadwin's hands were so badly blistered that he was unable to free her. He called to the truck driver for help and then he put his arms around Gray. "Don't die!" he begged. "Don't leave me behind!"

Hadwin had bundled her into heavy clothes and a sleeping bag for the long, cold trip, and according to her doctor, if not for this padding, she almost certainly would have died—either on impact or later, of exposure. Gray's ankles had to be repaired with screws and plates and now she must use a walker to get around. Hadwin visited her in the hospital every day until he left again for Haida Gwaii two weeks later, on January 12, 1997. "I've always wondered if Grant was trying to kill us both," she said, "so he wouldn't have to be alone."

ONCE ON HAIDA GWAII, Hadwin gave every impression of being a man on a one-way trip. While staying in a motel at the sparsely inhabited north end of Graham Island, he gave away all his possessions, including a number of things that had once belonged to his father. "Take whatever you want, because I'm going to burn the rest," he told Jennifer Wilson, the twenty-year-old daughter of the motel's manager. Hadwin went on at length about university-trained professionals, referring to them as "an incestuous breed of insidious manipulators." According to Wilson, he advocated terrorism as the most effective means of bringing about change, and he talked a great deal about trees. "I learned a lot from him about the forest," she recalled. "He seemed so passionate—like he wanted to do something good. I got the sense he had found his purpose." At one point she and Grant visited the golden spruce together. To a passerby, they might have made a pleasing and romantic picture: a handsome, youthful man and his attractive blonde companion—both of them clearly at home on the wild western rim of the continent. Grant was carrying a camera, and he asked Wilson to take a picture of him with the golden tree towering above. In his hand is a beaded eagle feather, a gift from a native elder.

After buying a gas can, falling wedges, and a top-of-the-line Stihl chain saw, Hadwin relocated to Port Clements, where he checked into the Golden Spruce motel. The last time Wilson saw him, he was wearing earplugs; he had to wear them, he told her, because every word he heard felt like a direct insult. Hadwin was traveling with medication, but it is pretty clear that by this stage he had either run out or simply stopped taking it.

Hadwin had replaced his totaled Honda, and during the night of January 20, 1997, he drove to the head of MacMillan Bloedel's Golden Spruce Trail. Having sealed his saw, wedges, gas and oil, and, presumably, his clothes into inflated garbage bags, he packed them down the short trail and descended the steep bank of the Yakoun. While it freezes over occasionally, the sixty-foot-wide river was open now and flowing more quickly than usual as it had been swollen by winter rains. The temperature was in the mid-thirties when Hadwin slipped into the current and swam across, probably one-handed, trailing his equipment behind him. The bank on the far side is equally steep and slippery, and it would have taken some doing to get himself and his gear up into the forest, particularly in pitch-darkness. There was no one around for miles, and as usual, clouds hovered low over the islands, enveloping everything in a miasm of mist and rain. Whatever light there was, Hadwin would have had to bring with him. The tree stood just back from the riverbank as it had for the past three centuries; it would have been a looming presence in the darkness, its golden qualities invisible in the sodden gloom.

HADWIN HAD CUT DOWN hundreds of trees in his life, but he had never tackled one this big; the golden spruce was more than 7 feet in diameter at its base, and it is rare to find trees this size around Gold Bridge, or elsewhere in the interior. By island standards, however, it was only of average size; there are specimens of Sitka spruce still standing on the coast which are 15 feet in diameter, but even larger ones have been reported in the Yakoun Valley. Like the redwood and redcedar, big Sitka spruce often grow "buttresses"—thick ridges that fan out from the trunk to help stabilize the tree on mountainsides and in shallow, rocky soil. In the early 1960s, a forester named Wally Pearson measured a spruce stump about ten miles upriver from the golden spruce that was 24 feet across—a diameter comparable to that of a giant sequoia; a 26-footer was reported further south in Sandspit. The only way to fell huge trees like this is to "dismantle" them: first by cutting the buttresses off and then by tunneling inward by cutting out what are called "window blocks."

In 1987 a Vancouver Island faller named Randy felled a red cedar more than 22 feet in diameter. Using a Husqvarna 160 chain saw with a 40-inch bar, it took him six and a half hours. After cutting a wedge all the way around the outside of the tree, he cut window blocks and tunneled into the center. The noise from his saw was so loud inside the chamber he had made, and the exhaust so thick, that he didn't know the tree was falling until daylight, let in by the lifting trunk, lit up the smoke around him. When a tree this big hits the ground, it doesn't sound like a tree; it sounds like a building collapsing at your feet. To those experiencing it for the first time, it gives new meaning to the expression "fear of God." When he examined the stump afterward, Randy recalled that "the [tree] rings were so tight you couldn't fit a piece of paper between them. That thing had to be—fuck—*thousands* of years old."

The response most fallers have to bringing down a massive specimen like this is similar to that of a hunter when he bags a trophy animal: it is at once beautiful, terrible—and immensely satisfying. But it is a rare occurrence nowadays; Randy's cedar was very likely the last of the truly huge coastal trees to be cut legally in B.C. "But even dropping the little ones—I still get a thrill," he added. "I'll never get tired of it. I've been hurt; I've had guys killed right next to me. But I guess that's why they pay us the way they do."

In B.C., a typical faller works a six-and-a-half hour day, for safety reasons, and until recently, a faller working for MacMillan Bloedel could make $800 (CND) a day. But since Weyerhaeuser took over, company fallers have been replaced, increasingly, by contract labor, with the result that day rates have dropped about 30 percent. Even so, it's still serious money for a job that appeals to only a tiny sliver of the general population. "Fallers are loners," explained Bill Weber, one of the few bullbuckers (falling supervisors) to survive the Weyerhaeuser takeover. "You're the master of your own destiny; you're not at the mercy of the machines. If you get whacked, you've got no one to blame but yourself."

There is no question that fallers, like high riggers, or helicopter pilots, are attuned differently than other people; you can see it in the way they move their heads. A conscientious faller will keep a close eye on the tree above him for the same reason—and with the same reflexive jerks of the head—that a feeding bird will: because in the forest, death generally comes from above. The top of the tree is where much of the information is; everything happening at the base is exaggerated at the extremities, much the way the end of a fishing rod will exaggerate the smallest movements of one's casting hand. Thus, the first indication that a tree is about to go can be seen in the quivering of the branch tips. These early vibrations can also shake loose dead branches, some of which can be the size of small trees; such deadly bolts from the blue are called "widowmakers." Should a faller happen to be hungover, looking straight up a long, gun-barrel trunk with its tapering perspective exploding into a kaleidoscopic spray of branches ten stories up will not only set his head spinning, it can induce nausea as well. Even when he's sober, the experience can be disorienting, and it is exacerbated by the tremendous amount of flex evident in a long trunk on the verge of falling. Unfortunately, this effect is easiest to see when mistakes have been made.

Standard procedure for falling a tree of any size, once the direction of fall has been chosen, is to cut a deep, grin-shaped wedge into that side of the tree; on the coast, this is called a Humboldt undercut. Occasionally a faller will cut his wedge too shallow or misjudge the tree's natural lean and, instead of falling over in the intended direction, the tree will settle back on the saw as it cuts through from the back side. This presents an extremely dangerous situation because it renders the saw useless and it means the faller is no longer in control. Now, the tree can do anything. Not only does it have 360 degrees of falling possibility, it can also kick out at the bottom, and even a glancing blow from that much mass will kill an elephant. Once a tree has settled back like this, falling wedges are the only way to persuade it to tip over in anything resembling a controlled fashion.

In addition to calk boots, many loggers today wear Kevlar pants (to mitigate chain saw accidents) held up with the logger's trademark red suspenders. Around their waist is a heavy leather belt with holsters and pouches containing chain saw tools, compression bandages, a cruising ax, and falling wedges made of high-impact plastic. A wedge, usually several, is inserted into the cut on the back side of a reluctant tree; some wedges are less than an inch thick, but in many cases, that is all it takes to shift a tree's equilibrium. At this point, the only things keeping the tree upright are a slender strip of holding wood, the downward pull of gravity, and the tree's own exquisitely balanced architecture. But a gust of wind or a snapping fiber can change all that in an instant, and as the wedges are driven home, a tree's flexibility becomes alarmingly apparent. The faller's rapid skyward glances will follow shock waves from each blow as they roll up the trunk, exiting through the topmost branches in rhythmic shivers. To some, this might seem gratuitously provocative—like kicking a giant in the shins—and they would be right. It takes a certain kind of person to bang on something wider than his front door, heavier than his whole house, and twenty stories tall when it's doing a snake dance.

Dennis Bendickson discovered early on that he wasn't that kind of person. Bendickson is a third-generation logger from a family that pioneered on Hardwicke Island. Silver-haired and solidly built, with forearms that still look powerful, he is now a senior instructor and program director in the Forestry Department at the University of British Columbia. Like most loggers, Bendickson went into the woods young; in his late teens he began falling big old growth. He knew he didn't have what it takes to keep at it when he asked himself, "Do I want to live to twenty-one?" "Cutting big trees, they usually need wedging to get them to go over," Bendickson explained. "I'd wedge and wedge, and they'd pop their holding wood; they'd be like a ballerina, spinning around on the stump. And you're down there running around like a squirrel trying to figure which way it's going to go, and you don't commit until the tree commits."

Referring to the academic joke about being "educated beyond your level of intelligence," Bendickson said, "There are some jobs where that's dangerous. I tried to think too much about what would happen; I'd try to work out the physics of it rather than rely on that sixth sense that good fallers seem to have."

This unmeasurable, nonintellectual awareness—what some call "bush sense"—is probably what keeps woodsmen like Randy, Bill Weber, and Grant Hadwin alive. Not only will a good faller have a better feel than most for how a tree will behave in a given situation, he may—like a gifted athlete—also have more "time" in the crucial moments to take in and process information and then determine the correct course of action—not by thinking, but by intuiting at a hyper- or extrasensory level (though dumb luck is a factor too). In the case of logging, the test for who has this gift, and who doesn't, is a terminal pass-fail. As Donnie Zapp, a Vancouver Island faller with thirty-five years in, put it, "It's not a job you want to bullshit your way into."

But chain saws make it look easier than it is. Of all the technological advances that have taken place in the forest, the most radical has been this one. Chain saws have been in development since at least 1905, when a two-man prototype was successfully tested in Eureka, California. After the logging frenzy of World War I, an amazing variety of devices was tested, but most of them, including one that used a red-hot wire to burn through trees, proved impractical in the rugged coastal forests. The chain saw as we know it today—essentially a motor-driven bicycle chain armed with sharp teeth—didn't become a common feature in the woods until after World War II.

Since then it has evolved into a sleek, light, and devastatingly efficient cleaver of the forest. Today, even a big chain saw like a Stihl 066 weighs under seventeen pounds; with a 36-inch blade turning at more than 60 mph, it has the same searing, attention-grabbing power as an AK-47 or a Gibson "Les Paul." Like a machine gun or an electric guitar, a chain saw is a handheld deus ex machina: a supercharged extension of masculine will that is impossible to ignore. They are thrilling tools to use. Some B.C. fallers, not content with stock performance, have their saw engines souped up to the point that their enlarged exhaust ports need spark arresters to prevent them from starting forest fires.

Under ideal conditions, chain saws function like noisy butter knives: one can buck up a large tree using only the weight of the saw and the pressure of one's trigger finger. But they will also take off a man's limbs as fast as a tree's. Given the right combination of opposing forces, they can behave like Ninja helicopters, and their tremendous power encourages a dangerously casual attitude toward smaller trees. A faller named Hal Beek discovered this in the worst way imaginable while working a setting on the west coast of Vancouver Island in 1998. Unlike second-growth tree plantations, which are usually monocultural groves all the same age, most old-growth forests contain trees from every stage of life; in between the giants are other aspirants of various sizes, including hundreds of saplings. As he travels from one big tree to the next, a faller will often use his saw like a slow-moving machete, swinging it back and forth in front of him— motor by the hip, blade angled toward the ground—to clear a path for himself. However, by cutting these smaller trees on a bevel rather than flat, the faller leaves a trail of "pig's ears"—pointed stumplets—behind him. Beek had cut a trail through a stand in order to get at a windfall cedar about six feet in diameter, and while standing atop the fallen trunk, he reached over and cut off another nearby sapling, leaving behind a pig's ear about five feet high. It was raining (as usual) and while Beek was bucking up the cedar, he slipped backward on some moss and impaled himself on this five-foot living spear; it entered through his rectum and didn't stop until it reached his spine. At that point, his toes were just touching the ground.

Fallers who have lost limbs to saws and shearing trees generally describe the experience as feeling like a "bump"; the real pain tends to come later. But an injury like Beek's is different; the pain he felt was instantaneous and indescribable. Every motion, even his attempts to call for help, would have been an agony unto itself the kind that would make most people pass out. Making matters worse was the fact that his legs were already fully extended: there was no way to free himself, and every movement risked driving the stake in further. Fallers generally work in pairs for safety reasons, and it is now customary for partners to call out to each other if they don't hear the other one's saw running, but Beek's partner was of the old school and he was oblivious; he heard neither Beek's shouts nor his emergency whistle. Beek realized that if he couldn't save himself, and quickly, he was going to bleed to death. Somehow he found it in himself to restart his saw, maneuver its three-foot bar behind him, and cut himself free—without amputating his feet, or collapsing back on the sapling or the saw. Then, with the five-foot stave still inside him, Beek crawled a hundred yards up an embankment, through heavy brush to a logging road. By the time the helicopter came, his friends were calling him Fudgsicle. After three months spent attached to a colostomy bag, and another three in rehab, he went back falling. This is not a unique occurrence; Beek's bullbucker, Matt Mooney, witnessed a similar situation in the Queen Charlottes when his partner fell on a broken branch; it entered by the same path and exited through the man's belly.

DESPITE HIS PROPENSITY for envelope-pushing, Hadwin was injured badly in the bush only once—when the pawl on a jack he was using slipped under load, causing the handle to flip up and shatter his jaw. The alarming frequency of accidents in the woods puts Hadwin's preference for working alone in a different light. Most responsible

companies wouldn't allow it now. Cutting down big trees in total
darkness is also frowned on.

The forest at night, in winter, is a very quiet place, and Hadwin's
saw would have sounded unbelievably loud in that peaceful setting. It
roared for hours, unheard, apparently, by all but Hadwin. The bullbucker for M&B who later performed what can only be described as
chain saw forensics on the tree, noted that Hadwin knew what he was
doing. He employed a Humboldt undercut and then cut a series of
"cookies"—small window blocks—to allow his 25-inch bar access to
the heart of the tree. He had clearly studied his target carefully
because he made his cuts and employed falling wedges in such a way
that the tree would not fall with its natural lean but rather in line with
the prevailing winds and toward the river. Sitka spruce is so strong that
two 30-foot logs connected by only four inches of heartwood can be
dragged through the forest without breaking, and Hadwin took advantage of this by leaving just enough holding wood so that the golden
spruce would remain standing until the next storm blew in.

But as Hadwin was making his cuts, he was—like every logger—
also carving his way into the past. Tree rings that had been hidden
since Harry Tingley picnicked there with his father, since the last
smallpox epidemic emptied the surrounding villages, since Captain
Kendrick was riddled with grapeshot, since a time before Captain
Pérez and Chief Koyah were born—all this fled by, unnoticed, in a
flickering comet's tail of sawdust. Hadwin didn't stop cutting until
about 1710, when his own ancestors were still living a near-tribal
existence in the British Isles and the masts of the first Nor'westman
had yet to puncture the southern horizon. Then Hadwin shut down
the saw, packed up his gear, and floated it back across the Yakoun,
leaving behind an audible silence and a tree so unstable that it would
have shivered with every breath.

THE NEXT DAY, Hadwin gave the saw away to an acquaintance
in Old Masset and caught a plane back to Prince Rupert. While he

was there, he stayed in the Moby Dick Inn, a high-rise motel three blocks from the water, and it was from here that he sent his final blast fax, copies of which were received by Greenpeace, Prince Rupert's *Daily News*, the *Vancouver Sun*, members of the Haida Nation, and even Cora Gray. But it was clear that the message was intended for another recipient: MacMillan Bloedel. It read, in part:

RE: *The Falling of Your "Pet Plant"*
Dear Sir or Madam:
. . . I didn't enjoy butchering, this magnificent old plant, but you apparently need a message and wake-up call, that even a university trained professional, should be able to understand.
. . . I meant no disrespect, to most of The Haida People, by my actions or to the natural environment, of Haida Gwaii. I do, however, mean this action, to be an expression, of my rage and hatred, towards university trained professionals and their extremist supporters, whose ideas, ethics, denials, part truths, attitudes, etc., appear to be responsible, for most of the abominations, towards amateur life on this planet.

A day later, the golden spruce came crashing down.

Locally, the reaction was overwhelming, particularly within the Haida community. "It was like a drive-by shooting in a small town," John Broadhead, a longtime resident of the islands explained. "People were crying; they were in shock; they felt enormous guilt for not protecting the tree better." Broadhead paused for a moment, trying to express the true impact in language someone who wasn't Haida would understand. "It was as if someone had done a drive-by on the Little Prince," he said at last. According to Haida legend, the golden spruce represented a good but defiant young boy who had been transformed, and because of this, some among the Haida saw the crime not as an act of vandalism, or protest, but as a kind of murder. "At a certain level it was real hurtful in the same way that New York [9/11] was," explained a Haida elder named Diane Brown. "A piece of our community was rubbed out."

As soon as they received the news, the Council of the Haida Nation issued the following press release:

> The Haida people are saddened and angered by the destruction of K'iid K'iyaas (Elder Spruce), also know as the "Golden Spruce," in the Yakoun River Valley on Haida Gwaii. The loss of K'iid K'iyaas is a deliberate violation of our cultural history. Our oral traditions about K'iid K'iyaas predate written history.
>
> We declare to the world that the Haida Nation takes full ownership of the remains of K'iid K'iyaas, and that it is declared off limits to everyone. The Haida will conduct a private ceremony at the site to reconcile the loss.
>
> The Haida expect that justice will prevail, and that the person responsible for the act of destruction will be punished. The Haida people will be watching every detail and if there is no apparent justice, the Haida will take appropriate action.
>
> . . . The Haida have long regarded K'iid K'iyaas as a sentinel of the Yakoun Valley, and now that it has been destroyed, the Haidas will escalate protectionist measures for our land.

For several days after leaving the islands, Hadwin remained in Prince Rupert at the Moby Dick Inn, where he stood out from the usual clientele, but not for the reasons one would have suspected. "There was a big difference between him and the fishermen and divers who come in here," recalled Pat Campbell, who worked the front desk. "He was more of an educated person, dressed sporty, neat and tidy."

Prince Rupert is literally the end of the line; it marks the mainland terminus of the transcontinental Yellowhead Highway, and once you are here, there is nowhere to go but out to sea, or back where you came from. The nearest town is eighty miles inland. Long the center of Canada's North Pacific fishing industry, Prince Rupert earned a reputation for speed- and cocaine-driven crews who would send tidal waves of cash washing through the bars, restaurants, and motels every time they hit town. It rains so much here that locals often don't bother wearing raincoats, and like most northern fishing communities, it has fallen on hard times. This is where one catches the ferries to Ketchikan and Haida Gwaii, and it was here that the Mounties caught up with Hadwin.

But they weren't the only people looking for him; Guujaaw, the future president of the Council of the Haida Nation, wanted a few words with him as well. Guujaaw, a singer, carver, activist, and politician, is one of the most powerful and charismatic figures in Haida Gwaii—a latter-day warrior. Descended from the legendary carver, boatbuilder, and storyteller Charles Edenshaw, he has about him the aristocratic air of a Balinese artist-priest, exuding the bone-deep confidence of one who is "to the manor born." Guujaaw managed to track Hadwin down before any deadline-bound journalist did, and the two men spoke on the phone. "He didn't seem crazy," Guujaaw recalled. "He sounded normal—neither excited, nor scared, nor regretful—as if what he'd done was no more than throwing a rock through a window. I asked him why he did it, and then I told him the story of the golden spruce and he said, 'I didn't know that.' He gave the impression that he probably wouldn't have cut it down if he'd known."

Philosophically, the two men weren't all that far apart; Guujaaw had been battling logging companies for twenty years, and as a result, he was sympathetic toward Hadwin's frustration. "He could have taken out a few [logging] machines; then he would have been respected," he said. In the end, though, Guujaaw compared Hadwin to John Lennon's killer: "a little man with, otherwise, nothing."

The Mounties visited Hadwin in person. After arresting him, charging him, and ordering him, to appear at the courthouse in Masset on April 22 (Earth Day), they released him on $500 bail. Already known to—and suspicious of—the police, he was offered no protection and did not request it. He was considered by some to be a flight risk, but there was so far no legal justification for keeping him in custody. Hadwin soon relocated to Cora Gray's home in Hazelton, 170 miles up the Skeena River, but his presence there caused other members of the Gitxsan tribe to fear that they would be seen as complicit in the crime, and they tried to distance themselves from the strange but generous white man in their midst.

Shortly after his arrest and release, the entire text of Hadwin's letter was published in the local papers, and for the next couple of weeks he carried on a dialogue with infuriated locals through newspapers on both sides of Hecate Strait. In an article entitled "Upset about the Golden Spruce? Re-examine your perspective, says Hadwin," he told a reporter for the Queen Charlotte Islands *Observer* that "we tend to focus on the individual trees like the Golden Spruce while the rest of the forests are being slaughtered." He then compared corporate set-asides like this and Vancouver Island's Cathedral Grove to circus sideshows. "Everybody's supposed to focus on that and forget all the damage behind it. When someone attacks one of these freaks you'd think it was a holocaust, but the real holocaust is somewhere else. Right now, people are focusing all their anger on me when they should focus it on the destruction going on around them."

While Hadwin did acknowledge his insult to the Haida, he fell short of a full apology. "There was no intent on my part to offend the native people in any way," he explained. "They should see a person who is normally very respectful of life and has done a very disrespectful thing and ask why."

But this was asking too much. Hadwin had cut down what may have been the only tree on the continent capable of uniting natives, loggers, and environmentalists, not to mention scientists, foresters, and ordinary citizens, in sorrow and outrage. Meanwhile, newspaper and television reporters from across Canada were flocking to the islands to cover the story, which also found its way into the *New York Times* and *National Geographic* and onto the Discovery Channel.

Scott Alexander, a spokesman for MacMillan Bloedel, was surprised by the flood of media attention: "This seems to have opened some kind of wound," he told one reporter. "I'm not sure why, but it's taking off more and more with each passing day." Cartoonists, poets, songwriters, and visual artists were also horrified and captivated by the death of the tree, and attempts to honor its memory were rendered in a variety of media that ran the gamut from doggerel verse to an exquisite Aubusson-style tapestry that would take a master weaver and her apprentice a full year to complete. In a few instances, these paeans veered off into uncharted territory: "Can there be another Golden Spruce?" lamented a columnist in Victoria's *Times Colonist*. "Can there be another Gandhi or Martin Luther King?"

"When society places so much value on one mutant tree and ignores what happens to the rest of the forest, it's not the person who points this out who should be labeled," Hadwin told a Prince Rupert reporter who questioned his sanity. In the short term at least, the collective reaction to the loss of the golden spruce ended up proving his point: that people fail to see the forest for the tree.

No one openly supported Hadwin, but there were those who sympathized with him. "I considered him misguided," explained one local logger, "but I could relate to his rationale—his hatred for M&B. Sometimes I'd just like to throw a bomb in their office." In an effort to explain how he was able to remain inside the industry under these circumstances, he said, "You don't allow yourself to think; if you start looking at it too hard, you're going to go crazy."

A young Haida man from Skidegate thought that what Hadwin had done was "a great idea. It was M&B's pet tree," he said, "but it's no more special than the thousands of others being cut down." In the past, he had worked as a logger when jobs were available. "You have the attitude," he explained, "that 'If I don't do it, somebody else will.'" Any of this man's ancestors who had been concerned about the declining otter population would have been driven toward the same logic and by exactly the same market forces.

HADWIN HAD BEEN CHARGED with felony criminal mischief —damage in excess of $5,000, and the illegal cutting of timber on Crown land. Ordinarily crimes like this draw a fine and/or minor jail time, but this was not an ordinary case and both the provincial authorities and the Ministry of Forests intended to prosecute to the fullest extent of the law. "He was going to get hammered in court," opined an RCMP officer from Masset named Blake Walkinshaw. "The courts are very lackadaisical in B.C.— from my point of view—but I think this one here was going to be an example." As an afterthought he added, "A person like that would have a hard time surviving in jail."

But this was a strange case; the law knew how to deal with timber poachers who cut down protected old-growth cedar, and it knew how to deal with arsonists who destroyed beloved cultural and historic sites, but how would it punish someone for cutting down a unique and sacred tree as an act of protest when most of the surrounding forest had already been felled for profit? On paper, Hadwin was facing years in jail and a heavy fine, but there was no precedent in Canada for how a local judge and jury might compute the far less tangible losses to the Haida, to the residents and economy of Port Clements, or to science.

There was, however, a precedent in Texas. It was set around the state capital's famed Treaty Oak, one of a group of trees known as the Council Oaks. Local Comanche Indians had once performed ceremonies within this sacred grove, and it was under the sole survivor that Stephen F. Austin, the founder of the state, allegedly signed the first border agreement between Indians and settlers. Once declared the most perfect specimen of a North American tree by the Forestry Association Hall of Fame for Trees, the 500-year-old live oak was poisoned in 1989 by a man named Paul Cullen; his motive, he claimed, was unrequited love. After extensive rescue efforts (financed with a blank check by the billionaire industrialist Ross Perot), a third of the tree was saved. Cullen was charged with a felony and sentenced to nine years in prison. Given that he tried to kill the Lone Star State's most venerable symbol, some might say that Cullen got off easy, and relatively speaking, he did: a life sentence had been seriously considered. No doubt alternative punishments had been contemplated as well, just as they were for Grant Hadwin, a man who some suspected wouldn't survive to see his court date. It was believed by the Mounties, as well as by local employees of the Ministry of Forests, that the Masset Haida might deal with Hadwin themselves. "A lot of problems are taken care of by the locals," explained Constable Walkinshaw. "That's why we don't have much trouble here. He might have been right [to fear for his life]."

One senior member of the Tsiij git'anee clan chose his words carefully, but did acknowledge that "unofficially, something could happen to him."

Masset is divided into two distinct communities: New Masset (population 950), the primarily Anglo village which includes the main shopping district as well as the federal dock and the courthouse; and Old Masset (population 700), the Haida reserve, which is, with the exception of non-Haida spouses, almost completely native. In addition to his more obvious crimes, Hadwin had disrupted the flow of life in this segregated community. "There's a rhythm to a small town," observed Constable Walkinshaw. "Old Masset, New Masset—everybody gets along quite well. Even Fran, the court stenographer, goes to the potlatches. Outsiders—people like Hadwin—put the rhythm out of sync. Somebody's going to get to him."

"Almost everyone in this community was ready to string that man up because of the hurt that was done to us," recalled Robin Brown, an elder from the Tsiij git'anee clan. "It was as if one of us had died." Ron Tranter, the Anglo resident of Old Masset to whom Hadwin had given his saw, was, briefly, a suspect in the crime, and he was furious. "If I see him," vowed Tranter, "I'll kill him." But there was, it seemed, a waiting list for this honor. "The consensus," claimed Eunice Sandberg, a bartender at Port Clements' Yakoun Inn, "is this guy should be done away with." A local logger named Morris Campbell suggested that they "nail his balls to the stump." One Haida leader also suggested that Hadwin be nailed to the tree; others were wondering "whether we should cut a part off the person who did this, to see how they like it." Talk is cheap, of course, but there is actually a kind of precedent for punishments of this kind; in *The Golden Bough*, Sir James Frazer's classic exploration of magic and religion, he wrote:

> How serious the worship of trees was in former times may be gathered from the ferocious penalty appointed by the old German laws for such as dared to peel the bark of a standing tree. The culprit's navel was to be cut out and nailed to the part of the tree he had peeled, and he was driven round and round the tree till all his guts were wound around its trunk. The intention of the punishment clearly was to replace the dead bark by a living substitute taken from the culprit; it was a life for a life, the life of a man for the life of a tree.

While the native residents of the islands have always been oriented toward the coast and the water, European settlers here have been inlanders—forest people—as much as they have been fishermen. Even today, the relationship many loggers have with the forest goes beyond simply cutting down trees. In this sense, not much has changed for a long, long time. It is hard for people outside this life to understand the logger's appreciation for his environment, but Jack Miller, who has spent sixty years in and around the logging industry, tried to demonstrate it with the following story.

Miller and his supervisor were cruising timber on Nootka Island, off the west coast of Vancouver Island, back in the fifties when his supervisor found an uncommon orchid and pointed it out. Miller soon found another one some distance away and said, "Here, I'll pick it for you."

But his supervisor told him to leave it where it was.

"Why?" asked Miller. "It's all going to be logged anyway."

"Leave it there," his supervisor ordered.

The individual's love of the woods exists in tandem with a collective industrial "rape and run" mentality that over time has left scoured valleys and fouled streams littered with machinery, fuel drums, old tires, and thousands of yards of rusting cable. Loggers, like most people who work for a living, see what they do as necessary. "It's a resource and it should be used" is a rationale one hears over and over again. But for most of the residents of Port Clements, what Hadwin did had little to do with resource use or environmental protest; like the Haida, they saw his act as the wanton destruction of a treasured symbol, a kind of sacrilege. Port Clements' mayor, Glen Beachy, echoed many of the Haida, when he told a reporter that "it makes me sick; it's like losing an old friend." But he had other things on his mind as well: "Why would a tour bus even come through here now?"

Mayor Beachy then declared a Random Acts of Kindness Week in Port Clements.

In an editorial, the managing editor of Prince Rupert's *Daily News* compared Hadwin's logic to that of the pro-life activist who would kill a doctor for performing abortions. By the end of January, feelings were running so high that the RCMP, under pressure to resolve the case as quickly as possible, moved Hadwin's court date ahead by more than two months. He was now to appear in Masset on February 18, just three weeks away. "They're making it as nasty as they possibly can," Hadwin told a reporter at the time. "They'll want me over there so the natives will have a shot. It would probably be suicide to go over there real quick."

And it may have been.

CHAPTER NINE

Myth

I will tell you something about stories
                    [he said]
   They aren't just entertainment.
        Don't be fooled.
   They are all we have, you see,
      all we have to fight off
         illness and death.

—Leslie Marmon Silko, CEREMONY

WHEN THE GOLDEN SPRUCE FELL, it knocked down every tree in its path. From a distance it looked like the wreckage left by a lightning strike, or a freak wind, which in a way, it was. After all, what were the chances? The golden spruce was one in a billion, and so was Grant Hadwin. "Whoever did this," said a MacMillan Bloedel spokesman shortly after the tree was found, "had to be hell bent." He was referring not just to the logistical details, but to the raw effort required to access the tree, and then to cut it down in the middle of the night. It is hard to imagine anyone else with the same combination of motive, obsession, endurance, and skill required to do such a thing.

The golden spruce fell in such a way that the last twenty feet or so hung out over the river, and it was a sorrowful sight: the still-luminous golden boughs thrown up like skirts, exposing the dark green underlayer; the sheared stump so startlingly white in the dark forest; the damage so small, relative to the great size of the tree, and yet so thoroughly irreparable. On Sunday, January 26, three days after the golden spruce was discovered by the wife of an M&B employee, the tree became the subject of a sermon in Masset's Anglican church. But it felt more like a eulogy. "This was not just a physical tree of unusual beauty," proclaimed the Reverend Peter Hamel, "it was in fact a unique symbol of the islands and ourselves. It was a mythic tree that sustained our spirits whenever we saw it. . . . The presence of this tree . . . brought us together and lifted us from the familiar to the divine." Hamel then called upon the great Romantic poet William Wordsworth to say what he could not:

> Oft have I stood
> Foot-bound uplooking at this lovely tree
> Beneath a frosty moon. The hemisphere
> Of magic fiction, verse of mine perhaps
> May never tread; but scarcely Spenser's self
> Could have more tranquil visions in his youth,
> More bright appearances could scarcely see
> Of human forms and superhuman powers,
> Than I beheld standing on winter nights
> Alone beneath this fairy work of earth.

"Confining the spiritual to the inner dimension of life," concluded the Reverend Hamel, "has given license to the violent exploitation of nature. The trees who clap their hands at God's justice suggest otherwise. All of reality is the realm of the spirit, of transforming upward encounter. The destruction of a tree, and in particular the golden spruce, has deep implications for us. This gift from Mother Earth connected us with our deepest spiritual needs. Its senseless destruction wounded each one of us as much as the loss of its wondrous beauty in the sacred grove by the Yakoun River."

The next day more than a hundred Haida traveled up the Yakoun in order to reconcile with the spirit of the golden spruce. "The elders were crying, praying in their own language," recalled a Haida clergy-woman named Marina Jones who attended the ceremony. "You could feel the heaviness; it was like losing one of our children. People were wearing their blankets inside out." Jones salvaged a golden twig from the tree; she had it freeze-dried and she keeps it in a hermetically sealed plastic package, like a piece of the True Cross. Urs and Gabriela Thomas, the proprietors of the nearby Golden Spruce Motel, keep a large golden sprig in a jar of alcohol by the reception desk, where it looks more like a specimen of rare coral than of a local tree.

John Broadhead, a director of a local environmental research group who has worked closely with the Haida for more than thirty years, put his finger on it when he said, "That tree was a lot more than a tree." Botanically speaking, the golden spruce was a mutant — a "freak" as Hadwin put it—but it was also the tip of a mythic ice-berg, and in this way it was a microcosm of the islands themselves. Some among the Haida refer to the Yakoun as the River of Life, and just as the islands seem to represent the life force in concentrated form, the golden spruce represented a concentrated essence of the Yakoun. In this sense, it has much in common with the more widely known concept of the Tree of Life. This ancient motif can be found today throughout the world—in Sri Lankan temples, Oriental car-pets, Middle Eastern and Meso-American ceramics, the Bible and even on bridge abutments in Southern California, among countless other places and media. The Tree of Life is a symbol of abundance, but it also represents a kind of metaphysical hub around which life and death, good and evil, man and nature, revolve in an endless cycle.

Vestiges of ancient rites related to trees can still be found in many parts of the world, including Europe, Africa, India, and the Far East; a few, like the maypole dance, the Christmas tree, and the Yule log, have survived the journey to the New World. But the memorial ceremony for the golden spruce may well have been the first of its kind ever to be held in North America. It is probable that nothing like it had been seen anywhere in the Northern Hemisphere since pre-Christian tribes worshiped in sacred groves, the same groves that were annihilated by invading Christian armies and governments—not just for the wood they contained, but for the pagan worldviews they represented. If one looks back far enough, it becomes clear that the Haidas' experience has been shared by almost everyone at one time or another. "Even now," wrote Pliny the Elder in the first century B.C.E., "simple country people dedicate a tree of exceptional height to a god. . . ."

On the following Saturday, February 1, a public memorial service to "mourn one of our ancestors" was held on the riverbank, opposite the fallen tree. It was raining—on the edge of snow—when the crowd came to fill the wound in the forest, and the Tsiij git'anee hereditary chief Dii'yuung entered the forest wearing a chilkat blanket woven of mountain goat hair and beating a black drum in a slow death-march cadence. Neil Carey, a U.S. Navy veteran and author who has lived in Haida Gwaii for fifty years, described the ceremony as "one of the largest collections of people from the islands I've ever seen in my life. It was just like a funeral; cars were lined up for a mile on both sides of the road."

It was Ernie "Big Eagle" Collison, also known as Skilay the Steersman, who organized and presided over much of the ceremony. With him were numerous Haida chiefs and other leaders, including Guujaaw, whose name means "Drum." At various times during the ceremony, the sound of drums grew so thunderous in the crepuscular silence of the forest that the beats seemed almost three-dimensional as they ricocheted off the tree trunks. Meanwhile, the singers' voices—Guujaaw's in particular—echoed through the woods as they belted out Haida songs of mourning and renewal. These were sounds that hadn't been heard on the banks of the Yakoun in ages, not since the great sickness, and the roar of chain saws that followed, had drowned them out.

The message sent by the chiefs, clergy, and community leaders was one of sorrow, forgiveness, and unity—but also of profound puzzlement: "It's hard to grasp how people think," said Chief Skidegate, who stood on a stage that had been built in the forest especially for this occasion and addressed the crowd through a microphone, "how someone could do such a tragic thing." When Ernie Collison took the stage he looked exhausted and bereft; ordinarily he was a vital man who rarely wanted for something to say, but now he seemed nearly at a loss for words. It was as if the death of the tree had sapped some of his own vitality. Collison described the ritual as a memorial ceremony to "address the feelings in your heart and in your soul and in your mind, in your fears, in your anger—with the loss of this beautiful tree that means so much to all of us on Haida Gwaii—Queen Charlotte Islands, if you will—and around the world. . . . People from all over North America have been calling," he said, "trying to put some understanding into the madness that created this sorrowful situation for us."

When a journalist asked Collison if he really believed a little boy could turn into a tree, he retorted, "Yeah. Do you believe a woman could turn into a block of salt?"

The Haida narrative canon has a lot in common with the Bible in that both contain stories that serve a variety of functions: some are creation myths; some keep track of family and tribal lineage; some are histories of the region and important local events; some are prophecies and others are told to instruct the young and remind the old. The golden spruce story, as it survives today, falls into the last category; it is a parable. All of these tales, properly told, carry important information as well as entertainment value, but much of this is lost in the twofold translation—first, from Haida into English, and second, from the spoken word to the printed page. Like a play or a song, stories of this kind were intended to be live events, energized by the charisma of the teller and by his or her connection to the audience. As is the case with Bible stories, literal readings of Haida narratives present problems for people seeking "rational," post-Enlightenment explanations. According to Haida legend, for example, Haida Gwaii is where the world began, and the first humans emerged from a clamshell at a place called Naikoon (Rose Spit), the long, sharply pointed sandbar at the northeast corner of Graham Island. So much of our understanding of "truth" and "fact" depends on context and orientation: to the uninitiated, the "Big Bang" theory sounds as bizarre and fanciful as the Haida story, "Spirit of the Atmosphere Who Had Himself Born." And yet, the former almost sounds like shorthand for the latter.

Since the Haida had no alphabet or writing, all information was passed down orally, and there were tremendous amounts of it; some Haida stories such as the creation myth "Raven Who Kept Walking" run for forty pages or more, but even at that length, it is almost certainly an abbreviated version of the original. This is not surprising when one considers the incalculable losses suffered by the Haida and their mainland neighbors. A resource map of the Queen Charlotte Islands, published in 1927 by British Columbia's Department of Lands, assessed the islands' timber holdings at over 15 billion board feet. It also indicates that virtually all of the prime forestland in the Yakoun Valley, including that around the golden spruce, had already been "alienated" (a British term meaning leased out). Most disturbing, though, is the map's tally of the islands' human population, which put the number of resident Haida at just 645. This startling figure represents a drop of approximately 95 percent from estimated pre-contact numbers (based on villages sites, shell middens, and other related data). Genocide, however passively it may have been perpetrated, is not too strong a word for this catastrophic decrease. Regardless of how one chooses to define it, the Haida very nearly went the way of the sea otter.

After successive waves of epidemics had reduced the Haida population to a skeleton crew, the survivors were in a state of shock, much as survivors of the Hiroshima bombing or the Rwandan massacres would have been. The mortally ill and rotting dead lay everywhere—too numerous to be moved or buried. Every aspect of the culture was effectively shattered, and the most basic activities ground to a halt. Too few people were left to paddle big canoes effectively, or fish, or tell stories, or to raise the orphans left behind; many of the skills and much of the knowledge died with its owners. It would be like going to work, to school, to a neighborhood bar, and finding nineteen out of every twenty people dead or dying with no help in sight. What do you do? Where do you go? By the turn of the century, survivors from approximately fifty villages—some of whom were bitter enemies—had been consolidated, first into five communities and then into two—Skidegate Mission, on the south end of Graham Island, and Old Massett on the north end. Even now, there are still sharp distinctions between, and within, them. "Talking about Skidegate and Massett is like talking about China and Japan," explained one elder who grew up in Masset. "Although the world has put us together, we know the difference." To this day, everyone knows who is descended from nobility, and whose ancestors were slaves.

The eradication of the culture continued as survivors were embraced by missionaries who helped to house, feed, and clothe their new charges but did so on Christian terms. Many Haida adopted this new faith, and it may have made a great deal of sense at the time, given the near total destruction of everything they had previously known and believed. "The island population is now shrunk to not over seven hundred," wrote the ethnographer and linguist John Swanton to his mentor, the famed anthropologist Franz Boas in 1901. "The missionary has suppressed all the dances and has been instrumental in having all the old houses destroyed—everything in short that makes life worth living."

In the course of the mass population losses and subsequent resettlement, nearly all the masks, costumes, and ritual objects that had once formed the material backbone of the Haida's spiritual life were lost. Some were abandoned or sold off by their owners who no longer had a context for using them and were desperate for cash to buy basic necessities. Others were gathered into piles and burned by missionaries, or confiscated by Indian agents who then sold the artifacts to collectors; anthropologists, too, carried off everything they could. By 1910, most of the poles that had stood on the northwest coast were gone as well: cut down under pressure from missionaries and government officials, or salvaged by collectors; some were cut up for firewood; in at least one case, they were used as pilings to hold up a waterfront boardwalk. A number of the best poles were taken to museums; in most cases, these salvage operations would turn out to be a blessing. But not all the Haida went quietly; at the south end of Prince of Wales Island (Alaska) were several villages of exiles known as the Kaigani Haida. There, a Chief Skowall co-opted the message of local Russian Orthodox missionaries by incorporating one of them, along with a Russian saint and Michael the Archangel, into a huge, new pole.

It must be said that missionaries, like all the other players on the coast, were a varied lot; some are remembered with great fondness and admiration for their generosity and guidance while others are sorely resented for their abuse and repressiveness. During the missionary period (1876–1940), a number of native traditions persisted by simply going underground. As a result, most Haida have a foot in both worlds; the Northwest Coast is home, now, to a diverse pantheon whose representatives range from the deserts of the Middle East to the Pacific Ocean floor.

In the decades following one of the most severe epidemics in 1862, many Haida left the islands to search for work, or simply for a place that was not so badly broken. Some made their way to Victoria, British Columbia's capital and first city, located at the south end of Vancouver Island. It was a 500-mile canoe journey for the Haida, and en route they were often harassed by tribes whose members they had stolen or murdered years, or decades, before. Once they arrived, Victoria did not treat them much better, and many Indians came to grief there. Before Vancouver surpassed it in the late nineteenth century, Victoria was B.C.'s center for logging operations. It is the only place on the continent where one can still see streets made from blocks of end-grain fir, laid together as bricks would be in any other city; the effect is of walking on a giant butcher block table. You would never know it today, but beneath the pretty flowers, elegant government buildings, and picture-perfect harbor are buried the roots of a rough outpost of empire inhabited by colonial government officials, loggers, sailors, Chinese and Hindu laborers—and traumatized Indians from up the coast. Alcohol, prostitution, and venereal disease followed each other closely within the destitute and demoralized native population, and it was not uncommon for former warriors to end up pimping their slaves, and even members of their own family, once they got to the city. In some cases, prostitution was even used as a revenue generator for funding potlatches, which had been held in secret ever since being outlawed by the Canadian government in 1884.

Still more nails were driven into the cultural coffin when generations of Haida children were taken from their homes and sent to residential schools where they were lumped in with children from other tribes, children like Cora Gray. Many of them did not know English, and native languages were forbidden. The government's objective was to take young Indians away from their "unimprovable" parents and turn them into Christian wage earners. To many whites, this no doubt seemed like a completely reasonable, even merciful thing to do. It was clear to them that the old ways were finished, and even if they could be revived, many aspects of that life—slavery, warfare, and raiding, to name a few—were untenable under the new regime. But the result of forced assimilation was, in many cases, an abysmal failure. In addition to being raped, beaten, and generally humiliated, generations of Indian children grew up profoundly disconnected from their families and culture and yet poorly equipped (not to mention unwelcome) to participate in the world of their conquerors. A number of these children, once released from residential school, also drifted south—first to Victoria and, later, to Vancouver—and many never made it back. This practice of what amounted to internment in residential schools began in eastern Canada nearly four hundred years ago and ended only in the 1970s. Legal claims against churches and the federal government for abuse suffered in these institutions now number in the tens of thousands.

The carriers of the culture tended to be those who slipped through the government agents' net and avoided residential school. These children stayed home with parents and grandparents; they learned the language, the stories, the skills, and a handful began the daunting process of patching together the tattered remains of their ancient legacy. Old Masset's residents had a coastwide reputation as carvers and canoe builders, and by the turn of the last century, they had somehow managed to retool and were turning out schooners and fishing boats as sleek and sturdy as any on the coast. By the 1940s there was a well-established fleet of Haida-built fishing boats plying the islands' waters. However, like fishermen and farmers today, these boats were financed with loans, often from the companies who bought their product. During the 1950s, much of the fleet fell into the hands of the fish companies due to unpaid debts, turning many Haida fishermen into hired hands on their own boats. While the art of canoe building has since been revived, the Masset Haida lost the art of modern boat building and, along with it, control of their economic destiny. However, they regained something else that may, in the end, prove more important: their stories and ceremonies—the core of the culture.

The curiosity—and the courage—to revisit these losses did not return for decades. An extraordinary recovery process began in the 1960s when Haida artists began to revive the lost arts of pole carving, mask making, and canoe building. With extensive assistance from a number of committed individuals and organizations from off-island, the Haida have performed a monumental feat of self-reclamation that combined mining the memories of the elders with visiting museums throughout the world in order to reacquaint themselves with all that had been lost, stolen, and sold away during the nineteenth century. And it is ongoing; early films and field recordings made by anthropologists aid them in recalling their songs and dances, and ancestral bones are now being repatriated from museums and given proper burials; artifacts, too, are being returned. The scattered pieces of what was very nearly a lost tribe are gradually returning home.

In 1969 Masset's first post-missionary pole was carved by Robert Davidson, who was one of the leaders of the Haida renaissance. Davidson's grandmother wanted to dance at the pole raising, but nothing like this had happened on Haida Gwaii for generations. Lacking anything in the way of the masks or costumes once associated with this activity, she wore a paper bag over her head. It was like a scene from *Fahrenheit 451*, this elderly woman—one of the last links to a culture that had been thousands of years in the making— shuffling across the floor, leading the others as their feet rediscovered the lost steps, and the words reassembled in their minds and mouths to resonate once again after a terrifying silence. It was this generation—the one that had personally known the smallpox survivors— who saw to it that the golden spruce story, along with so many others, survived to the present day.

ON THE NORTHWEST COAST, stories are considered property, just as land or automobiles are in Euro-American culture, or as guitar tunings are in some Hawaiian families. Some stories are held in common while others belong to a specific clan, or family, who are the only ones entitled to tell it; the same goes for certain dances, songs, and heraldic crests. If you were to ask a Haida in Skidegate to tell you the story of the golden spruce, she would say, "That's not our story," and then she would send you up to Masset, several days' paddle— now, an hour's drive—to the north. If you were to ask a Masset Haida to tell you the story, he might tell you, if he knew it, but he would probably refer you to an elder in the Tsiij git'anee clan.

The story of K'iid K'iyaas—the golden spruce—was not committed to paper until 1988, less than a decade before the tree was cut down. The carriers of the story, who had learned it in the Haida language, were frustrated repeatedly as they tried to pass it on to Caroline Abrahams, a Haida teenager who collected the story for inclusion in a book on the history of the Yakoun River. Abrahams, who now lives in West Virginia, is, like all of her peers, unable to speak or understand the Haida language (there are fewer than forty fluent speakers left, and the youngest of them, Diane Brown, is in her fifties; the rest are twenty years older or more). The elders who recounted the story—among them, Abrahams's grandmother—stopped repeatedly in mid-sentence to say, "There is no word for this in English." As is the case with the adoption of any conqueror's language, it is more an act of necessity than a labor of love and the result is often a less-than-perfect grasp. Thus, to read a native Haida speaker's English version of the golden spruce story is kind of like reading one of *The Canterbury Tales* in rudimentary Haida. It is impossible to calculate just how much nuance, meaning, and art remain on the far side of the translation, but it is considerable. Still, like all good stories, the story of K'iid K'iyaas has some universal qualities; it combines elements of the stories of Sodom and Gomorrah and Noah's Ark as well as the Greek myths of Artemis and of Orpheus and Eurydice. But this version begins with a boy shitting on the beach.

One winter many years ago, a young man went down to the beach to relieve himself. It was too cold to squat so he stood. When he was finished he looked down and there was his turd, standing straight up like a tree in the snow. The young man thought this was funny and he laughed and laughed. And that is when the snow began to fall without stopping. All the winter supplies ran out, and still it snowed. One by one the villagers died of cold and starvation until only two people were left: an old man and his grandson. They realized their only hope lay in trying to escape the doomed village, and so, with the blizzard still raging, they dug their way out. Once they had traveled for some distance, they were amazed to find the forest alive with summer.

As they walked, searching for a new home, the old man issued a warning. "Don't look back," he said to the boy. "If you do, you will go into the next world. People will be able to admire you, but they won't be able to talk to you. You'll be standing there until the end of the world."

However, the way was long and tiring, and the boy missed his fishing gear. He couldn't resist stealing one last look at the only home he'd ever known. But when he did, his feet took root in the forest floor. The boy cried out for help, but despite his grandfather's best efforts, he remained rooted in the ground. "It's all right, my son," said the boy's grandfather. "Even the last generation will look at you and remember your story."

It was this boy who became the golden spruce. There are stories up and down the coast of rocks, islands, and mountains representing humans, animals, and spirits who have been transformed; even Vancouver has Siwash Rock, a fifty-foot-high sandstone pillar representing another disobedient boy who was transformed after defying the gods. But of all the known Haida or West Coast transformations, the golden spruce is the only one that involved a living* creature who could be seen by everybody whether they were native or foreigner, believer or skeptic. The golden spruce, in fact, was uniquely suited to bridge the gaps of time and culture. Trees are the only readily visible living things with such tremendous temporal reach, and no other tree was so strangely distinctive, so undeniably Other, that it could be recognized instantly by anyone, no matter what their culture, or at what point in history they came upon it. Left in peace, the golden spruce could have lived until the twenty-sixth century. The stump revealed not a trace of rot, even though some internal decay is common in coastal trees more than 250 years old.

In addition to this version of the story about the making of the Haidas' golden boy, there are variations. One tells us that the snow

---

* Implied here is a scientific definition of "living," as opposed to that traditionally used by most native peoples, which held that everything was alive and interconnected. Increasingly the world of science is subscribing to this view as well.

came as a punishment from the creator for intratribal fighting; another attributes this freezing flood to a general lack of respect for nature, demonstrated symbolically by the boy laughing at his own feces. Another version describes the two lead characters as the sole survivors of a smallpox epidemic who wanted to live forever. Yet another take on this tale maintains that the tree would live as long as the Haida Nation, and that its death would herald the end of the tribe. "Regardless of how you state it," observed the elder Robin Brown, "people are going to contradict you." While people who have grown up with the written word might view these variations as "inconsistencies," it is worth remembering that prior to the publication of the first English dictionaries in the seventeenth century, even spelling was a highly subjective business; each rendering of a word was the result of an individual's personal decision in the moment. Oral traditions are not so different; each version of a story is highly dependent on a given teller's memory, integrity, agenda, and intended audience, but it also depends on the current needs of the teller, the listeners, and the times.

At the root of the golden spruce story, though, is a very simple message: respect your elders, or you'll be sorry. However, beneath this surface layer of meaning, the parable could also be read as a lesson on how to survive the loss of one's entire village to a massacre or smallpox or, for that matter, how to weather a stint in residential school: don't look back; don't try to return to that dead place. But everyone in a position to deny or confirm this, or any other theory, is dead. Even the grandmother who heard this story as a girl and passed it on to her granddaughter has passed on. Like the tree and the man who cut it down, the story is a puzzle or, more accurately, a piece of a puzzle, the whole of which can never be fully known.

CHAPTER TEN

Hecate Strait

What's that he said—Ahab beware of Ahab—
there's something there!

— Herman Melville, MOBY-DICK

AFTER A BRIEF STAY in Hazelton, Hadwin returned to Prince Rupert in order to prepare for his trip to court. While Cora Gray was shocked by what he had done, she remained loyal to him. "He did wrong," she told a journalist at the time. "He feels bad about what he done. He could only see MacMillan Bloedel. He didn't see no legend about the Haida when he did that." Gray went so far as to try to book Hadwin a room for his upcoming stay in Haida Gwaii. According to her, everyone she spoke to there said they had no vacancies, though this is rarely the case, even in high summer. Given that Gray was calling in darkest February, it is more likely that no one wanted Hadwin under his roof. By this time Hadwin's options were becoming stark and few: he could face the music, or he could run. His situation would be most people's idea of a nightmare, but for Hadwin it may well have provided a kind of golden opportunity. For the first time in his life, he had an awful lot of people's attention and, given his convictions and his previous willingness to go on the record, there is every reason to suppose that he saw the courtroom as an excellent forum in which to air his grievances. He just had to get there in one piece. Hadwin's solution to this problem, like his solution to MacMillan Bloedel's timber practices, was filtered through a complex—and to most people baffling—mix of pride, personal integrity, paranoia, and absolute conviction. In this sense, he was not so different from people like Joan of Arc or Ted Kaczinski; he even had a certain charisma, though like the Unabomber, he lacked the ability to persuade and inspire. However, there were two crucial ways in which Hadwin differed from these other committed, radicalized, and egocentric individuals: first, he was neither a killer nor an advocate of killing, and second, he had unassailable credentials in the area of wilderness survival. It was this unwavering self-confidence in the face of the elements that led Hadwin to attempt something no one had before: a mid-winter crossing of Hecate Strait by kayak. There are compelling reasons why this has never been tried, and Pat Campbell, who was working the desk at the Moby Dick Inn, where Hadwin stayed, summed them up pretty well: "The water in Rupert is boiling, rough water," she explained, "and that's just by the dock. I'm sure [Hadwin] would have known. He would have seen the weather—what it can do; it's vicious that water, just vicious."

A NUMBER OF PLACES lay claim to the title "Graveyard of the Pacific," and the west coast of Vancouver Island is one of them, but it would be more accurate if its limits were extended to include all of coastal B.C. Well over a thousand vessels have gone down here during the past two hundred years, and Hecate Strait is arguably the most dangerous body of water on the coast. The strait is a malevolent weather factory; on a regular basis its unique combination of wind, tide, shoals, and shallows produce a kind of destructive synergy that has few parallels elsewhere in nature. From the northeast come katabatic winds generated by cold air rushing down from the mountains and funneling, wind-tunnel style, through the region's many fjords, the largest of these being Portland Inlet, which empties into the strait thirty miles north of Prince Rupert. Winter storms, meanwhile, are generally driven by Arctic low pressure systems born over Alaska, and they tend to manifest themselves as southerlies along the coast. It is because of these winds that the weather buoy at the south end of Hecate Strait has registered waves over 100 feet high. One of the things that makes the strait so dangerous is that these two opposing weather systems can occur simultaneously. Thus, when a southwesterly sea storm, blowing at 50 to 100 miles an hour collides, head-on, with a northeasterly katabatic wind blowing at similar strength, the result is a kind of atmospheric hammer-and-anvil effect. Veteran North Coast kayakers tell stories of winds like this lifting 400 pounds of boat and paddler completely out of the water and heaving them through the air.

But this is only one ingredient in Hecate Strait's chaos formula. Tides are another; in this area they run to 24 feet, which means that twice each day, vast quantities of water are being pumped in and out of the coast's maze of inlets, fjords, and channels. The transfer of such volumes in the open ocean is a relatively orderly process, but when it occurs within a confined area like Hecate Strait that is not only narrow but shallow, the effect is of a giant thumb being pressed over the end of an even larger garden hose. The scientific name for this is the Venturi effect, and the result is a dramatic increase in pressure and flow. A third ingredient is a frightening thing called an overfall which occurs when wind and tide are moving rapidly in opposite directions. Overfalls are steep, closely packed, unpredictable waves capable—even at a modest height of 10 to 15 feet—of rolling a fishing boat and driving it into the sea bottom. They can show up anywhere, but their effects are intensified by sandbars and shoals like the one that extends for twenty miles off the end of Rose Spit between

Masset and Prince Rupert. Under certain conditions, overfalls take the form of "blind rollers," which are large, nearly vertical waves that roll without breaking; not only are these waves virtually silent, but under poor light conditions they are also invisible—until you are inside them. If one then factors in the prevailing deep-sea swell that in winter surges eastward through Dixon Entrance at heights of 30 to 60 feet, and the fact that a large enough wave will expose the sea floor of Hecate Strait, the result is one of the most diabolically hostile environments that wind, sea, and land are capable of conjuring up.

Most sailors who survive storms do so because they orient themselves to the prevailing wind and waves, get into the flow, as horrendous as it may be, and ride it out. But on a bad day in Hecate Strait, you can't get into the flow because there is no flow to be found; a 70-knot gust or an apartment building's worth of water can hit you from any direction. There is no rhyme or reason; all around you, the elements are at war with themselves. Because of the manic-depressive weather and the fact that this part of the coast remains as dark and featureless as it was when Pérez came through, mariners must navigate these waters the same way a mouse negotiates a kitchen patrolled by cats: by darting furtively from one hiding place to the next. If the conditions aren't favorable, you simply sit tight and wait—maybe for a long time. As one local veteran put it, "The worst thing you can do is be in a hurry to get somewhere."

Gordon Pincock is an expert kayaker and one of the people who pioneered the sport in Haida Gwaii. Over the course of twenty years he has paddled the length and breadth of the archipelago, including numerous trips along the extremely exposed and isolated west coast. On one occasion, he survived a day in 30-foot storm swells, during which he was nearly killed by a West Coast phenomenon called clapitos. Clapitos occurs when a large wave bounces off a cliff face and collides with the wave behind it, turning the sea into an aqueous trash compactor. It is hell on small craft: a 30-foot wave ricocheting off a wall will head back to sea as a 15-footer, but when it butts heads with the next 30-footer the two will merge into a 40-foot mountain of confused hydropower—over and over again. It is significant, then, that Pincock has never attempted a crossing of Hecate Strait. "Go out there alone?" he said. "In February? No way! I would never risk my life doing that, not even in the summer."

WITH HIS BELONGINGS LIQUIDATED and his safety in doubt, Hadwin was down to the contents of a single suitcase and a Visa card. Among the last items charged were a sea kayak, emergency flares, two paddles, and a bailing pump—standard equipment for a paddling trip on the Northwest Coast. Hadwin's stated destination was Masset, and he was leaving in good time to make his court date. He had told people that he was traveling this way because he was afraid he would be attacked by locals if he took the ferry or a plane, and he had legitimate grounds for concern. "People were going to see that he didn't get on the ferry," explained Constable John Rosario, who handled the case at the Masset end. "The feeling in Masset was that there was going to be a lynching if he came back."

But on this matter Hadwin seemed both lucid and resolved; shortly before he left, he telephoned the Haida leadership and announced his intentions: if they wanted to, he said, they could meet him out on the water where there would be "no uniforms [police] around." After notifying Cora Gray, his estranged wife, Margaret, and the *Daily News*, Hadwin launched his kayak on the afternoon of February 11. Both Gray and his wife notified the Mounties, who dispatched an inflatable powerboat and intercepted Hadwin as he was leaving Prince Rupert Harbour. But Constable Bruce Jeffrey, one of the officers on the scene, was unable to dissuade him from going. "He wasn't irrational," recalled Jeffrey. "He wasn't suicidal, but I could tell he was a few fries short of a Happy Meal. Unfortunately, you can't arrest someone for being overconfident or foolish. If he'd said, 'I'm not going,' we'd have flown him over, but he was determined to go under his own steam."

At dusk, with his gear stowed in fore and aft compartments and an ax and a spare paddle lashed to his forward deck, Hadwin paddled out of Prince Rupert Harbour and directly into a storm. Weather reports for that night show breaking waves over 10 feet, winds gusting to more than 30 miles an hour, and rain. Keeping one's bearings along this anonymous coast is difficult even in broad daylight, but it is impossible at night, in those conditions. It would have been so dark that even white-capped waves would have been barely visible. The temperature was just above freezing, but the windchill factor would have driven it down to zero; under these circumstances, an ordinary person would be at risk for frostbite within half an hour. Hadwin was wearing only a slicker and dishwashing gloves; he was not an experienced kayaker, but even if he had been, it was unlikely that he could have survived a night in such weather. And yet, somehow, he did. Sometime around midnight, he found his way back to Prince Rupert.

"He was waiting at the door when we opened," recalled Marilyn Baldwin, who co-owns SeaSport, where Hadwin had bought his kayak and equipment the previous day. Baldwin remembers Hadwin seeming surprised at how cold he had gotten in the night; he told her that he had paddled for hours in heavy seas and had been unable to make any headway. He could handle the breakers, he said, but he had returned to buy some warmer clothes and (on Constable Jeffrey's advice) a chart for Hecate Strait. When the topic of the tree came up, "he wanted to argue," recalled Baldwin. "I think he wanted his day in court. He got very agitated; his muscles were *vibrating*—like something taut, ready to snap."

Baldwin wasn't sure if this was because he was stressed or hypothermic, but as soon as he left, she phoned the police. However, Hadwin was acting within his rights, they said; there was nothing they could do. At dawn on the thirteenth, with five days left to make his court date, Hadwin set off again. This time, he didn't come back.

Clothes notwithstanding, Hadwin had equipped himself well for the task at hand; his kayak was a Nimbus "Telkwa," a high-end model made from laminated strips of Kevlar and fiberglass that is designed to carry heavy loads on long trips in rough conditions. In addition to being considerably longer than a river kayak, a sea kayak like the Telkwa has a V-bottom which enables it to track across the wind, rather than get blown sideways like a beer can or a raft. It is also equipped with a foot-operated rudder which allows a kayaker to devote all his paddling energy to driving the boat forward rather than steering. Hadwin's kayak was eighteen feet long, and while big, heavy boats like this offer more stability in rough seas, they also present more surface area to crosswinds and waves, which will tend to grab the bow and push the boat off course. Even though a kayak's low center of gravity can be a great asset in bad weather, there is no getting around the fact that they are small and fragile craft; a 3-foot wave, properly timed and shaped, can flip one with no trouble.

Marilyn Baldwin, like Constable Jeffrey, felt confident that Hadwin didn't have a death wish. Hadwin had told a journalist that he could do the trip in twenty-four hours, implying a nonstop crossing of Hecate Strait. Baldwin, however, was under the impression that he knew what he was up against in attempting such a trip, and that rather than heading due west and directly into the overfall zone off Rose Spit, he would island-hop, as the Haida once did. Such a route would have taken him in a northwesterly arc up and over to Prince of Wales Island, or perhaps even further west to Cape Muzon at the south end of Dall Island. From there, he would still have to sprint the last forty miles across Dixon Entrance, but, in taking this longer route, he stood a better chance, both of traveling in a following sea and of avoiding overfalls. Under ideal conditions, this last leg would take close to twenty-four hours on its own, but ideal conditions don't present themselves in Dixon Entrance in the month of February, especially in total darkness.

The following morning, February 14, a white kayak identical to Hadwin's was sighted off Port Simpson, twenty-five miles north of Prince Rupert. It was almost certainly him because no one else would have been out paddling in such conditions. The wind was out of the south that day, and gusting into the thirties; the waves rolling in from Dixon Entrance were pushing 15 feet. This was no kind of weather for kayaking, but Hadwin had the wind at his back and he was making excellent time—the question was, where? To a casual observer—and there were several that morning—Hadwin appeared to be bound for Alaska, but this is also the route a cautious (a relative term here) kayaker might take if he was island-hopping to Masset. Port Simpson marks the southern entrance to Portland Inlet, which traces the U.S. border. However, the 25-mile run from there to Cape Fox on the American side is locally notorious: in addition to being a point of collision for katabatic outflows and inbound southerlies, the tides here can reach 5 knots—riptide speed—which will defeat the efforts of the strongest paddler. Furthermore, when southerly winds like those blowing at Hadwin's back hit an outgoing tide, the inlet's mouth is churned into what local boatmen call "river chop"—steep, sloppy waves that are, essentially, aspiring overfalls. Sometimes they seem to defy the laws of physics: imagine breaking waves a dozen feet high but only eight feet apart. "We can't haul in that stuff," explained a Prince Rupert tugboat captain named Perry Boyle. Boyle's biggest tugboat has 1,200 horsepower and weighs 100 tons; Hadwin's kayak, in comparison, might as well have been a Popsicle stick powered by a goldfish. While there are definite advantages to being light and maneuverable, even in bad weather, they may well have been outweighed by the continuous exposure to wind and waves that a journey like Hadwin's would have entailed—no matter where he was going. The moon was at half and waxing, which meant successively larger tides with each passing day, and the high winds and low barometric pressure that accompanied the storm systems now pulsing through the strait would have made the tides even higher than normal. Over the next four days the weather would deteriorate steadily.

HADWIN HAD FORCED his way into the local consciousness barely three weeks earlier and yet he had already acquired a quasi-mythical aura; like Billy the Kid or the Scarlet Pimpernel, he seemed capable of turning up anywhere at any time. Even though there had been no sign of him—at sea, or on land—in four days, many residents of Haida Gwaii fully expected the killer of the golden spruce to appear in the Masset courthouse at 9:30 A.M. on February 18. No one seemed too concerned about the weather that morning, despite the fact that out in the strait, gale-force winds were driving horizontal rain through a cloud ceiling you could just about reach up and touch with your hand. Once again, the islands were concealed; had Captains Pérez, Cook, Vancouver, or Dixon been searching for land that morning, they would have sailed right on by. Hadwin may have had trouble finding the islands too, and not simply because of poor visibility; in Dixon Entrance, the seas were mounting to 30 feet.

The Masset courthouse is located inland in the center of New Masset on a side street lined with low shops; playing fields stretch away to the northeast and beyond them lies a sprawling plywood warren of a building that houses the rec center. Three miles to the northwest, along a narrow beachfront road, lies Old Masset, the Haida reserve. The courthouse itself is a postmodern ziggurat of aluminum and plate glass whose interior is defined by white linoleum and fluorescent lighting; it is one of the Crown's most remote and modern outposts. Canada is a constitutional monarchy, and here a courthouse doubles as a temple to colonialism. to this day, one is charged not just for violating the criminal code, but for disturbing "THE PEACE OF OUR LADY THE QUEEN HER CROWN AND DIGNITY." However, standing in front of the Masset courthouse while ravens clack like ratchets as they rifle through the garbage cans and the feeble morning light fights a losing battle with a North Pacific gale, the queen, not to mention the capitals of Victoria and Ottawa, might as well be a million miles away. Even so, the rule of law appeared to prevail; the anticipated lynch mob was nowhere to be seen. Instead, a line of people in heavy coats and hats were queued up at the courthouse door, backs to the wind. Metal detectors aren't a regular feature of life on Haida Gwaii, but on this day everyone was getting scanned. Inside, the hallway and waiting room would soon be packed to capacity, with the overflow spilling outside. Meanwhile, in the small courtroom, made smaller still by the dense crowd and damp, close air, the sense of anticipation was palpable. A cross section of the islands was in attendance: chiefs and elders, loggers and fishermen, housewives and shopkeepers sat in stiff rows on wooden benches waiting to lay eyes on the man many felt had attacked each of them personally.

Because of the islands' remoteness, there is no resident judge; instead, a provincial judge is flown in once a month to hear cases. For this reason there were, packed in among the people waiting for Hadwin, other islanders who also had hearings scheduled for that day. Normally these would be discreet, semiprivate affairs, but this morning, those accused of stealing an outboard motor, or driving drunk, found themselves under the scrutiny of nearly a quarter of Masset's adult population. It was embarrassing and a bit surreal— like a parody of Norman Rockwell's painting *The Four Freedoms*.

Thomas Grant Hadwin was called at nine-thirty, and there was a collective indraft of breath as a hundred eyes swept the room. Since few islanders knew what he looked like, most people weren't quite sure who they were looking for beyond some unfamiliar movement, the face of a stranger, or some vision of the man they carried in their head. In the end, no new face or energy field presented itself; the room remained the same, so everyone ended up looking at each other while the five syllables of his name simply hung in the air. What had begun as a haiku ended as a koan. Even after it was clear that Hadwin wasn't in the building, no one left; everyone waited, wondering where he might be: was he in custody, in hiding, on the run, dead— or just late? Hadwin's name was called again at ten, and this time someone stood up in the aisle. For a brief moment some in the room thought it might be Hadwin, but it wasn't; it was a Gitxsan man named James Sterritt who claimed that Hadwin had retained him as his agent. They had agreed to meet on his court date, Sterritt said, but he hadn't heard from Hadwin in two weeks. When the judge asked him if he had been authorized to act on Hadwin's behalf, Sterritt allowed that he hadn't. It was at this moment that Hadwin officially became a fugitive.

WHEN HADWIN'S ESTRANGED WIFE first heard that Grant had gone missing, she wasn't all that concerned; he had disappeared before, and apparently he hadn't always been truthful about where he was going. Armed, now, with another warrant for his arrest, the RCMP were taking more of an interest in this character, particularly when his wife described him as "indestructible." The test of his well-being, claimed Margaret, who had many reasons to be skeptical, would be whether he called their daughter on her birthday. Hadwin may have been in big trouble, but he was still a father, and in his own unusual way, a loyal one. When March 1 came and went with no phone call, Hadwin's wife began to fear the worst, and the Canadian Coast Guard began searching in earnest; meanwhile, U.S. authorities had also been put on alert.

For some in the U.S. Coast Guard, there may have been a strange sense of déjà vu; after all, they had searched for Hadwin once before. In the spring of 1993, while on his paranoia-induced sojourn through the north country, he took an open-ended side trip to Alaska's Alexander Archipelago, which lies about 200 miles north of Haida Gwaii; though tightly packed into the fragmented coastline, it looks like a mirror image of its Canadian counterpart. Hadwin landed in Sitka, the former capital of Russian America; once a major fur-trading station, it remains one of the most beautifully sited communities on the coast. In addition to being the origin of the Sitka spruce tree's name, the fort town was the site of the greatest single massacre of the fur trade era. Redoubt St. Michael, as Sitka's predecessor was known, was built on Tlingit territory, and in 1802 warriors clad in animal-headed helmets and armor attacked the town, killing four hundred inhabitants and enslaving the rest. Only a handful of people escaped. Two years later the Russians retook the site with the help of ship's cannon. As out of the way as it appears now, the settlement was once known as "the Paris of the Pacific;" for the first half of the nineteenth century, it was the most important port on the West Coast.

Shortly after he had arrived there, Hadwin rented a kayak from the chairman of Greenpeace's Alaska chapter; he planned to go paddling for a week but ended up being away for more than two. When he failed to return on the date indicated in his float plan, an extensive search was launched involving Coast Guard vessels and aircraft, local police and state troopers, and a volunteer search-and-rescue team. The first thing they found was Hadwin's abandoned campsite at the foot of an enormous snowcapped volcano on the south coast of Kruzof Island, which lies west of Sitka, on the outer edge of the archipelago. The camp looked as if Hadwin had simply walked away, leaving behind his tent, cooking gear, kayak paddle, spray skirt, and numerous other items. Bears were plentiful in the area so searchers immediately wondered if he had been attacked, but rescue dogs found no sign of him. Meanwhile, an aerial search discovered Hadwin's kayak floating upside down near St. Lazaria Island, a small bird sanctuary off the south end of Kruzof. Hadwin's backpack was strapped to the aft deck and inside it they found what they believed was a suicide note. However, on closer inspection it turned out to be something far more unusual.

Ordinarily Coast Guard search-and-rescue reports are rigidly formulaic—dense with the highly technical shorthand of aviators, mariners, and meteorologists. At the end of these forms is a place for "Remarks," and it was here that the officer charged with writing up the Hadwin report broke out of role for a moment and scrawled, "THIS WAS A DOOZIE." When you turn the page, you see what he means; this is where Hadwin takes over. The appended document is entitled "THE JUDGMENT"; it is fifteen pages long and impeccably typed. Considering it was written by a high school dropout who had felt compelled to leave first his country and finally his campsite because he believed he was under surveillance by the CIA, the contents are surprisingly cogent and considered. But what is more remarkable is that, unlike the writers of most manifestos, rants, screeds, and religious harangues, Hadwin begins, not by telling the reader what to think, but by asking him a series of questions, in effect, employing the Socratic method to put the reader in the Creator's shoes.

"I ASK YOU," begins the introduction,

If you had the power, to create all matter, including life, and you could synchronize, those creations, perfectly, what would you do, if one life form, was apparently abusing, all other life, including themselves?

If the original "INTENT," of your creation, had apparently been twisted, from "RESPECT," to hatred, from compassion, to oppression, from generosity, to greed, and from dignity, to defilement, what would you do?

How would you convince, people, that material temptations, social status, and education institutions, are used, to preserve and perpetuate, the status quo, with very little *real*, considera- tion, for the future, of life, on earth?

. . . How would you, as "THE CREATOR OF LIFE," show your *con- tempt* and *revulsion*, for the institutions, and the individuals, who are supposed to protect life, but are apparently, doing something quite different, instead?

Hadwin then goes on to give a brief history of the world, focusing on the transition from hunter-gatherer to settled agrarian, and from there to our current dependence on global commerce. He stops along the way to offer a thoughtful analysis of how relations between the "female nurturer" and the "male hunter, killer, gatherer and provider" conspire to encourage degradation of the environment. And he takes great pains to outline our progressive disconnection from nature and its negative effects, both on human beings and on the planet. Not only is this development contrary to the Creator's will, writes Hadwin, it is undemocratic:

A democratic society is morally responsible, for the actions of its institutions and elected or appointed representatives, at home or away. It is the responsibility of all individuals, in democratic societies, to resist any crimes against life or suspected crimes against life. Ignorance, abuse or lack of physical presence, at the scene of the crime, are not necessarily valid excuses, unless there are severe, mitigating circumstances. . . .

Finally, Hadwin outlines a radical solution to what he sees as a world gone horribly wrong: dismantle society as we know it, abolish all currency and religion, and remove all men from power. Replace the status quo with small, agrarian villages run by women and restricted to pre-industrial technology. The sole purpose of these matriarchal communities would be to repair the damage wrought by the past two thousand years of male-dominated civilization. It should be noted that Hadwin's was a hyper-masculine world in which women played very traditional, housebound roles; his wife was a quiet, dedicated homemaker who cooked from scratch and kept close watch over the children. His mother, too, was proud of her support position (even when addressing children, she would introduce Tom Hadwin as "my husband the engineer"). That Hadwin would wish to fire his own gender from all positions of authority is unusual, and his decision to forgo apocalyptic vengeance—a staple of most cosmic housecleaning scenarios—is equally radical.

It is in this unique and strangely appealing document that Hadwin the forest-loving woodsman and Hadwin the conscientiously objecting visionary merge and integrate. Paul Harris-Jones, the timber cruiser turned forest rescuer, and Professor Simard, the forest tech turned researcher and educator, had similar awakenings, as have countless others. But the big difference between most of them and Hadwin is one of intensity and context. Hadwin wrote that he had a spiritual experience on a mountain near McBride, B.C., during which he was not only forgiven for his prior sins but chosen to represent the

Creator of all Life and carry a message to the rest of humanity. An event like this goes by different names depending on when and where it occurs. One or two thousand years ago it would have been called a vision or a revelation, and the person who claimed it might be ignored like a fool, revered like a god, or killed like a heretic— sometimes all of the above. In more recent times, many of those who have entered religious orders were not hired or head-hunted, but called as by a voice, hence the term "avocation." Nowadays someone who gets blindsided by such a sudden and mind-altering experience might call it an epiphany, an awakening, or a religious experience while a professional might call it a delusion, a hallucination, or a psychotic episode. The truth is often somewhere in the elusive middle, and yet billions of people continue to be guided in their lives by just such liminal figures, most of whom—like Jesus, Buddha, Muhammad, and Brigham Young—are long and safely dead. Were they alive today, they might be languishing in a heavily medicated limbo, or, if they were lucky, they might be sent to Dr. Lukoff.

Dr. David Lukoff is a psychologist who has taught at Harvard and UCLA and who is now at the Saybrook Institute in San Francisco; he has made a specialty of treating people with stories like Hadwin's. In so doing, he has also coined what may prove to be a more useful term for these cataclysmic personal events; Lukoff calls them "spiritual emergencies." During a spiritual emergency, one is often granted access to what Michael Harner, a well-known anthropologist and expert on shamanism, calls "non-ordinary reality." While most of us find the idea of such experiences alarming, shamans actively seek them out. Wade Davis, the noted ethnobotanist, once commented to a journalist that "I never met a shaman who isn't somewhat psychotic—that's his job." Like Harner and Davis, Lukoff is intimately acquainted with this neighboring universe because he has spent time there himself; in fact, Lukoff had an experience that bore striking similarities to Hadwin's—right down to the benign, planet-repairing utopia he envisioned and the compulsion to write down his vision and disseminate it to the world (Hadwin's "Judgement" has been distributed on at least three continents). It took Lukoff, then in his twenties, months to regain his equilibrium and realize that his was not a central place in the universe but rather one among many. However, it was this experience, and its painful fallout, that allowed him to hear his avocation, which was to help the latter-day Ezekiels, St. Anthonys, and Hildegard von Bingens who are swept, unprepared, into states of searing, otherworldly awareness.

In 1985 Lukoff proposed adding a new category to the American Psychiatric Association's *Diagnostic and Statistical Manual* (*DSM*); he wanted to call it "Mystical Experience with Psychotic Features." One reason Lukoff and his colleagues were pushing for this addition was because recent surveys were revealing some surprising data about psychiatric patients and the people who treat them: despite the fact that nearly three-quarters of patients surveyed indicated that they had at some time addressed religious or spiritual issues in treatment, and two-thirds of them used religious language when discussing their experiences, fully 100 percent of surveyed clinicians indicated they had received no education or training in religious or spiritual issues during their formal internship. When Hadwin was interviewed at the forensic hospital in Kamloops, these findings were borne out; while one doctor noted that Hadwin saw himself as having a "special role" in the world, another simply determined that he had "very overvalued ideas about the environment and fighting the establishment." This is a decidedly sinister assessment: how, one might well ask, is it possible to "overvalue" air and water? Perhaps a truer indication of mental illness (or, at least, psychospiritual disconnection) can be found in the far more common tendency to passively accept the abuse of the very systems that keep us alive. In any case, this experience might explain some of Hadwin's hostility toward "university trained professionals." As Lukoff wrote, "Ignorance, countertransference, and lack of skill can impede the untrained psychologist's ethical provision of therapeutic services to clients who present with spiritual problems."

Gene Runtz, the man who hired Hadwin to do road layout up in McBride in 1987, was probably the first person to encounter Hadwin after what was likely his first "spiritual emergency." It was Runtz who compared his star contractor's frightening transformation to that of Dr. Jekyll and Mr. Hyde. To those who knew him during this period, Hadwin's blistering conviction sometimes came across as messianic—which, undoubtedly it was—but as disturbing (or laughable) as such pretensions might seem, they are par for the course. Roberto Assagiolo, a pioneering Italian psychologist who specialized in the relationship between psychology and spirituality, was well acquainted with the delusions of grandeur that often accompany spiritual emergencies. "Instances of such confusion," he wrote in a landmark paper entitled "Self-Realization and Psychological Disturbances," "are not uncommon among people who become dazzled by contact with truths too great or energies too powerful for their mental capacities to grasp and their personality to assimilate." Had Joan of Arc or Muhammad ibn Abd al-Wahhab (the founder of Wahhabism) been sent to see Drs. Assagiolo or Lukoff, the history of Europe and the Middle East might have unfolded in very different ways. Or perhaps not; the catch with this type of individual is that when he is still reeling from his encounter with the powers beyond, he is almost impossible to reason with. This may explain why so many people thus afflicted end up living in caves, on mountaintops, or on remote islands in small, sympathetic communities.

In 1994, a year after Hadwin's round-the-world trip and his subsequent psychiatric evaluation, the APA published the latest edition of its manual (*DSM-IV*) in which it included Lukoff's proposed category, though under the more generic heading of "Religious or Spiritual Problem." Four years later Lukoff founded the Spiritual Emergency Resource Center, a clearinghouse for information about the phenomenology and treatment of such emergencies, including case histories of people who have successfully integrated these experiences.

THE AMERICANS' SEARCH for Hadwin at the south end of Kruzof Island continued for three more days, but finding no sign of him, the Coast Guard called it off and contacted his next of kin. However, after learning from his wife that Hadwin was a skilled outdoorsman who could "survive on nuts and berries for six weeks at a time," they suspected he might still be alive, and three days later they resumed searching. Four days after that, a fishing boat reported smoke from the northwest coast of Kruzof, more than twenty miles from Hadwin's original campsite. A Coast Guard vessel was dispatched only to be met with a lukewarm reception by the man they had been trying to find for more than a week. According to the Coast Guard report, "This person's attitude left us with the impression that he really didn't care to be found."

If Hadwin's intention was to get away from it all, retiring to a windswept clifftop overlooking the North Pacific with a sleeping volcano at his back was a pretty good way to do it. The desert fathers would have approved, even if Search and Rescue didn't. Why Hadwin chose such an exposed location is a mystery; it might have been the place he felt the most direct connection to the creator, or it may have been a way of testing himself. Perhaps the roar of the wind and surf drowned out the racket in his head, just as earplugs would silence the rest of the world before he cut down the golden spruce. Then again, it may have been the only place outside of a tent where he wouldn't get eaten alive by mosquitoes. June is a bad month for bugs in Alaska; generally it takes a good five or ten knots of breeze to keep them at bay, but even then they will tend to hover in your lee, waiting for the wind to die. Mosquitoes swarm so thickly up there that they can, like clouds, briefly assume recognizable shapes. This is probably the only circumstance in nature where it is possible to look downwind and see a shadow of oneself infused with one's own blood.

According to the Coast Guard report, Hadwin had been living on mussels and clams for "many days"; his unwelcome rescuers noted that the bushes behind his campsite were littered with shells. He claimed he had beached his kayak above the high-tide line at the south end of the island and then decided to "do a walkabout." Shortly afterward a major storm blew through and it was this, apparently, that set his kayak adrift. Guessing his boat would be long gone after such a blow, he hadn't bothered going back for it. Hadwin had no sleeping bag and so had been living in the open ever since abandoning his campsite ten days earlier. Despite the storm and nighttime temperatures that dipped into the thirties, he was warm, dry, and in fine health. All he had besides the clothes on his back was a plastic bag of matches and some coffee.

CHAPTER ELEVEN

The Search

For there is hope of a tree, if it be cut down, that it will sprout again, and that the tender branch thereof will not cease. . . . But man dieth, and wasteth away: yea, man giveth up the ghost, and where is he?
—Book of Job 14:7–10

HADWIN WOULD BE much harder to find the second time, but during the months following his disappearance there were some tantalizing glimpses. Someone thought they saw him down in Bella Bella, an extremely remote native community on one of the coastal islands where a character like Hadwin would have stood out like a sore thumb. Someone else thought they saw him up in Hyder at the north end of Portland Canal, an equally remote community of Anglos where Hadwin would have fit right in—and did. Hyder is a wormhole in the U.S. Canadian border, a little piece of Alaska that can only be accessed via water, or B.C. Highway 37-A; it is a place where, as one Alaska state trooper put it, "everyone is suspect." Hadwin stayed there several times in 1996 during his extended travels through the north country. The rough dirt road that runs through the middle of town ultimately dead-ends at a glacier field, and this may have been one reason Hadwin liked the place. He also may have been drawn to its atmosphere of vigorous defiance; over the past thirty years Hyder's U.S. and Canadian customs houses have been shot at, burned down, and generally harassed by Indians and Anglos alike. At one point, the Canadian border post was subject to a psyops campaign consisting of a PA system blasting a continuous loop of "North to Alaska." Wedged tightly between coastal mountains, Hyder began life as a hiding place where Nisga'a Indians are said to have sought refuge from marauding Haida; after several mining booms during the twentieth century, it has once again become a sanctuary. Today, the town has no police force and is home to about one hundred people who count privacy and discretion among their highest priorities—kind of like the residents of Gold Bridge. Both communities share a common prayer:

> Dear Lord, give us one more boom.
> We promise not to piss this one away.

It was in March that a second warrant for Hadwin's arrest was issued—this time by the RCMP in Stewart, just across the border from Hyder. Hadwin had been ordered to appear in court there following his accident on the Nass River bridge—the charge: "driving without due care and attention"—but now that court date, too, had come and gone. The search for Hadwin was a daunting task that was pursued with no great enthusiasm; he hadn't killed anybody, and there was no reward for his capture or recovery. Besides, it was hard to know where to look; in between Hyder and Bella Bella lie literally hundreds of islands and thousands of miles of coastline backed by dense forest. An army could hide in there—a dozen armies, and yet of all the islands that dot the coast here, there was one that kept attracting attention.

Mary Island is about four miles long and lies about seventy miles northwest of Prince Rupert, at the entrance to Revillagigedo Channel and the northern portion of the Alaska Marine Highway System, which runs from Bellingham, Washington, up the B.C. coast to Skagway and Haines. Also known as the Inside Passage, this is the main thoroughfare for northern coastal traffic. Ferries, barges, and fishing boats thread their way through this network of tight channels by the hundreds all year round; in summer they are joined by fleets of cruise ships and sailboats. As busy as it sounds, it is still an anonymous place—less a destination than a way to somewhere else; the area is virtually uninhabited and there are no places for large vessels to stop between Prince Rupert and Ketchikan. There are, however, lots of places to hide a kayak; if Hadwin had been making a run for the border, this is almost certainly the route he would have taken.

A month after Hadwin's Masset court date, a man was found stranded on Mary Island, and his reasons for being there were unclear. No one had reported him missing and nobody seemed to know anything about his trip. He claimed his inflatable skiff had overturned in rough weather while en route from Ketchikan to Hyder, and that he had been stranded for three days during which he had survived on mussels and stream water; he was extremely hungry. The man, who had suffered frostbite on his feet, bore some resemblance to Hadwin, even sharing the same birth year; his name, allegedly, was Dennis Harrington, though Dennis Roe was also a possibility. The details were at once close enough and sketchy enough that the RCMP was contacted, but in the end they concluded that Harrington/Roe was a different person, possibly the survivor of a shipwreck involving large bags of marijuana that were discovered floating in the surrounding waters. At around the same time, on a mainland beach adjacent to Mary Island, the top of a skull turned up; it had a hole in it of the kind a bullet might have made. The fragment was taken to a medical examiner who determined that it was too old to be Hadwin's. It was so old, in fact, that it could conceivably have been a remnant from the same encounter described by Richard of Middle-giti'ns, the only participant in a Haida canoe battle who was ever formally interviewed.

Richard was born about 1850 on Chaatl Island, which lies at the entrance to Skidegate Channel, just off the southwest coast of Graham Island. A member of the Pebble-Town people, he lived for a time in Alaska and worked for the Hudson's Bay Company before returning to the islands, where he spent his last years in Skidegate Mission. This is where the American ethnologist John R. Swanton met and interviewed him during the winter of 1900–1901. While in the employ of the American Museum of Natural History and the U.S. government's Bureau of American Ethnology, Swanton did extensive ethnographic research on Indian tribes throughout North America, and he is credited with gathering most of the surviving pre-contact Haida history and mythology. He also collected the following account, and the distance from this to Homer's *Iliad* is not a long one.

Sometime around 1870, Richard participated in a raid on a party of Tlingit, a mainland Alaska tribe who were bitter enemies of the Haida. It was a revenge mission, and in order to settle the score, Richard and his party of two canoes filled with warriors traveled more than 150 miles up the eastern shore of Graham Island and across Dixon Entrance, to reach Tlingit territory. They wore knives hung from lanyards about their necks, and when it came time to fight they would tie their knives to their hands. They also carried spears and rifles, with cartridge boxes tied to their waists; a shaman went with them and one of his duties was to "whip the people's souls" before battle.

Somewhere in the vicinity of Revillagigedo Channel the Haida encountered a group of women and armed Tlingit warriors in a canoe "so large that the people in it could not be counted." When the Tlingit saw them they paddled away, firing two shots as they retreated, one of which killed Richard's brother. The Haida returned fire and killed the Tlingit steersman; then they shot two more Tlingit braves. The Tlingit fired back, grazing Richard's skull; then they motioned for the Haida to stop. The Tlingit didn't want to fight anymore, but one of the Haida canoes pursued them. As they closed in, a Tlingit warrior stood up and threatened to shoot at them, but a Haida killed him with a bone spear. "He dropped the gun," recounts Richard. "The Tlingit then quickly sat down. He pulled out the spear. His intestines came out at the same time. He broke it. And when he started to shove the spear back into the wound, someone [from the Haida canoe] jumped in to him." The canoe Richard was in then joined the battle, now a full-blown knife fight.

While killing a number of Tlingit—including one "whose insides fell upon me"—Richard was stabbed in the shoulder, "which made my insides come together [with pain]." After the battle had been going for some time, "a youth having no knife then made with his hands the motion of surrender to me from the bow. And I picked him up and threw him into [our] canoe. When another came at me I struck him. It grazed him. He went at once into our canoe. He let himself be enslaved. I made a cut down his back. He was a brave man. [This man was apparently a notorious Tlingit chief named Yan.] When it was reported that he had let himself be enslaved the Tlingit became boneless [with astonishment]."

[Afterward] a Tlingit was lying upon one of our young men. And, pushing away his knife, I cut off his head. . . . I looked toward the stern and they were already taking slaves. And when I went thither I saw one woman left. She had been shot in one leg. And I did not take her. The property was captured at once. Into [the other] canoe they took ten severed heads. There were only nine slaves. And after SKA'ngwai's father had brought five heads into ours, they found fault. He stopped then. And they took all the property.

In front of the place whence we had been wrangling a whale swam about with its young one. And we shot the young one. We killed the young one. We took its oil to Port Simpson to trade. There we bought all kinds of stuff. . . .

The warriors now got in. And, as they went along, they

began to sing war songs. It was hard for me. Two of my
younger brothers were killed, and I sang differently from
them.

The victorious warriors, some of whom were grievously injured,
then recrossed Dixon Entrance, whereupon they encountered a band
of Masset Haida who harassed them for attacking the Tlingit. The
Massets tried to take away the Pebble-Towners' recently acquired
slaves, and this led to a fight between members of the two Haida
bands, but it was broken up. Eventually they shared an awkward
meal, during which neither side laid down their weapons. Tobacco
was then brought out and enormous offers of blankets and firearms
were made for the prize captive, Chief Yan, but the Pebble-Town people refused to sell. Despite the fact that many of Richard's paternal
family lived in Masset, the situation remained tense. "We remained
awake that night. A part of us slept ashore. I was all covered with
blood from fighting." In the morning they paddled for home, passing the village of Skidegate, whose inhabitants were known for
intercepting other Haida raiding parties and stealing their slaves.
But on this occasion they stayed ashore. "After having fought we
sang songs of victory for many nights," concludes Richard. "Here is
all of this story."

Because these kinds of skirmishes were commonplace, there is no
telling whose head the medical examiner was looking at, but it
underscores a problem encountered by anyone attempting to do
forensic police work in this part of the world: the Northwest Coast is
very hard on evidence. Sergeant Randy McPherron is a homicide
detective with the Alaska state troopers who handled the Hadwin
case on the American side, and the challenges he faces on the job
are radically different from those confronting his urban counterparts. "Alaska's a good place to get rid of a body," he explained,
"especially in Southeast [the Alaska Panhandle]. People died all
over the place up here; the area was heavily populated with natives,
and there are hordes of unsolved murders. A lot of times it's a nee-

dle in a haystack." Half the battle is getting to the evidence before nature does: "There are so many things that can happen up here," said McPherron, "so many kinds of wildlife that will dispose of bodies."

Bears will do the heavy lifting while mice, seabirds, eagles, and ravens will pick over the bones. Crabs and insects will take care of the rest. Between the rain and the scavengers, the window of forensic opportunity might be only a matter of days—if that. "If he did roll over and sink," said McPherron, "nothing will ever turn up. There's a lot of deep water out there, and in cold water, when something sinks it stays down."

Regardless of whether a body sinks and becomes a "submarine," in marine rescue parlance, or rises again to become a "sailboat," anthropophagy, the technical term for the consumption of a human, will begin almost immediately. Because sea life is so abundant along the coast, an aggressive combination of shrimp, sea lice, dogfish, and crabs can skeletonize a body within twenty four hours. This is why victims of drowning are seldom recovered here. If they do turn up, it is generally because they have sunk in shallower or warmer water where the gases generated by internal decomposition carry them back to the surface. Should a body sink particularly deeply into an anaerobic zone devoid of plants and other sea life, it may become adipose, a condition in which subcutaneous fat reconstitutes into a kind of sealant, sometimes called "mortician's wax." Under these circumstances, a body can remain intact almost indefinitely; lost divers have been found in these waters a decade after disappearing, their neoprene-clad corpses still prowling the sea bottom like PADI-certified headless horsemen.

On April 4, the Prince Rupert RCMP received a request for Hadwin's dental records, but the records didn't match. Five days later, missing person posters were distributed up and down the coast and, on April 12, a patrol flight searched the coastal area north of Prince Rupert again. Hadwin had now been gone for two months, and at this point there was a lull; the posters elicited no new information and no new searches were ordered. Many assumed that Hadwin had either drowned or fled the country, and it seemed now that, one way or another, the golden spruce had been avenged. Meanwhile, some extraordinary things were happening in Haida Gwaii.

Ernie Collison once described K'iid K'iyaas as "a perpetual tree," and among the Tsiij git'anee elders there were some who claimed that the golden spruce was not the first such tree to grow in that spot, that it had been preceded by another golden spruce. This is another one of those stories that is hard to explain in a "rational" way, and it calls into question the relationship between the story and our relatively modern—and linear—concept of time. Perhaps this wasn't a story in the sense of its having already happened and thus being confined to the past; perhaps it was a story in the Haida sense in which time operates more like a spiral, or like the rings of a tree. There is a saying among the peoples of the Northwest Coast: "The world is as sharp as the edge of a knife," and Robert Davidson, the man responsible for carving Masset's first post-missionary pole, imagines this edge as a circle. "If you live on the edge of the circle," he explained in a documentary film, "that is the present moment. What's inside is knowledge, experience: the past. What's outside has yet to be experienced. The knife's edge is so fine that you can live either in the past or in the future. The real trick," says Davidson, "is to live on the edge." This is where Davidson, the most famous living Haida artist, spends most of his time, and this may be where one needs to be in order to fully apprehend the story of the golden spruce. The idea of there being more than one golden spruce on that site by the Yakoun may have been as much a version of history as it was a version of the future, for now it seemed that the golden spruce might rise again.

The Haida were not aware of this, but more than thirty years earlier, myth and science had been set on a collision course, and on January 25, 1997, they met inside the head of an Englishman named Bruce Macdonald. That morning, Macdonald, the director of the University of British Columbia's botanical garden, read an alarming headline on the front page of the *Vancouver Sun*: LEGENDARY TREE'S LOSS SICKENS RESIDENTS.

The University of British Columbia occupies a broad swath of prime real estate five miles west of downtown Vancouver, at the end of a high peninsula. The university's 110-acre botanical garden lies at the edge of campus, where it offers more than ten thousand plant species and commanding views of Georgia Strait, Vancouver Island, and Washington State's Olympic Mountains. As Macdonald made his way through the *Sun* article on that unusually clear and icy morning, his mind flashed to a shady patch of sloping forest in the garden's native plant section. Growing there was a pair of small Sitka spruce trees whose needles had a peculiar tendency to turn golden yellow. Even so, they were easy to miss in such a grand setting, and because they grew in the shade, their golden qualities were patchy at best; this, combined with a predisposition to grow sideways, gave the trees a hunched, anemic appearance. To the untrained eye, they would have seemed like prime candidates for culling.

Macdonald knew nothing of the trees' provenance as he had been living in England when they were first planted, but he went immediately to the garden's accession records to see if they might be related to the tree he had read about in the paper. UBC accession number 18012-0358-1978 is described as a "golden Sitka spruce" (*Picea sitchensis* 'Aurea'); the source was the Queen Charlotte Islands. They had to be from the same tree. The specimens, it turned out, were the all-but-forgotten legacies of three men: Gordon Bentham, an amateur conifer enthusiast, Oscar Sziklai, a Hungarian plant geneticist, and Roy Taylor, a former president of the Chicago Horticultural Society and director of the Chicago Botanic Garden, who had been Macdonald's predecessor at UBC. In 1968, the same year he became the director of UBC's botanical garden, Taylor copublished a two-volume 800-page survey of the flora of the Queen Charlotte Islands. Taylor's books made no mention of the golden spruce, but he was well aware of the tree and hoped to acquire a specimen for UBC. He would end up with two of them, but it would take more than ten years. As it turned out, the golden spruce was extremely difficult to reproduce.

Since the early sixties, senior foresters at MacMillan Bloedel had been hoping to propagate the golden spruce for the company's arboretum on Vancouver Island. Their interest in it coincided with a period of new and aggressive research into tree breeding and propagation as M&B sought to develop plantations of "elite" Douglas fir that were selectively bred from the finest wild specimens. In order to do this, the company had retained Oscar Sziklai, a pioneer in the field of tree breeding and one of 250 students and professors from the forestry school at Sopron, Hungary who immigrated to Canada en masse following the unsuccessful Revolution of 1956. With support from H.R. Macmillan, they ended up being hosted by UBC's nascent forestry program, where Sziklai became a full professor. Throughout his career, he participated in scientific collaborations and exchanges across Europe and Asia, and in 1986 he became the first foreign member of the Chinese Society of Forestry. Sziklai's original interest in the golden spruce lay in the question of whether its golden quality could be passed down genetically. However, closer inspection revealed that the tree was sterile; it produced very few cones, and none of their seeds appeared to be viable. This detail is consistent with a version of the Haida story which claims there were two golden spruces and that the second tree was a "male," unable to reproduce.

Shortly before he died in 1998, Dr. Sziklai told a journalist that, on one of his many visits to the golden spruce, "a Haida princess guided us to the tree and said, 'If the tree dies, the Haida Nation will die.'" At the time, Sziklai was a prominent scientist under contract with the country's biggest lumber company, and yet after thirty years, he still remembered this encounter; it may have been another reason he took such an interest in the tree. There was no guarantee that attempts to clone the tree would be successful, but if it could be done, Sziklai was the man to do it. He was given the job on one condition: that he would keep his findings secret. "I wasn't allowed to come to the public loudly and say, 'We can propogate it,'"he told a reporter, shortly after the tree was felled. "They guarded this tree closely to the heart, and felt the public would strip the tree and it would disappear."

"If we'd publicized it," said Grant Ainscough, a former chief forester for M&B, "we'd have had nothing left but a stump."

In the 1960s, the propagation of West Coast timber species by artificial means wasn't all that well understood, and no research was being done into Sitka spruce because it wasn't a commercial priority at the time. The preferred method of propagation was to take cuttings—scions—from a desirable tree and either graft them to other rootstock or plant them directly. Either way, it was a crude process that generally began with a blast from a hunting rifle because the easiest way to take cuttings from big trees is to shoot them off. Professor Sziklai, in particular, was known for his expert marksmanship; armed with a pump-action Remington, he was able to knock down individual cones from hundreds of feet away. What propagators didn't know forty years ago was that each part of a spruce interprets its genetic and hormonal instructions literally. Like a dog, the older a Sitka spruce gets, the harder it is to teach it new tricks, and like a member of a rigidly structured caste system, a branch never forgets its place in the pecking order. Thus, if your scion comes from a lower branch growing out of a trunk as old as the golden spruce, it will continue trying to fulfill its mission of being that branch, even when it is grafted to different rootstock, or planted vertically. Eventually it was discovered that branches closest to the apex of the tree were more willing to adapt to a new role—that of an upward-growing leader or trunk (which is what propagators generally want). It is because of this that when the top of a hemlock, cedar, or spruce gets blown off, you will often see the topmost surviving branches bending upward to replace the lost leader, giving the tree the appearance of an enormous candelabrum.

Sziklai chose to treat his cuttings with a rooting hormone and "set" them directly into soil, rather than graft them, and the results were discouraging. Of the two dozen cuttings planted, only half of them took root, and from then on their prospects went from bad to worse. According to a MacMillan Bloedel newsletter from 1974, despite "meticulous care and attention," only three of these original cuttings retained "their golden tones," and none of them grew at a normal rate. "Nature," it said, "appears reluctant to duplicate a rare, beautiful mistake." (One of these golden clones was presented in secret to H.R. MacMillan himself, but it died not long after.) Although Sziklai made more than one attempt, only a very few of them survived; despite being close to forty years old, the most vigorous of these is less than 20 feet tall and the only reason it is growing vertically is because the tree was bound to a stake for its first ten years. Something was clearly missing; moisture and cloud cover were obvious guesses, as most of the cuttings were planted in southern B.C., but there may also have been some more elusive, perhaps ineffable ingredient.

CHAPTER TWELVE

The Secret

Gray, dear friend, is all theory,
And green is the golden tree of life.
　　　　　—Johann Wolfgang von Goethe, FAUST

FROM THE POINT OF VIEW of physics, all of us
are rebels because we spend our lives actively subverting the forces
of gravity and entropy, two of the fundamental laws to which all
earthly matter must ultimately answer. But the tree is the greatest
living symbol of this twofold defiance. Trees are simultaneously
photo- and geotropic; that is, they are programmed not only to seek
out the shortest path to the noonday sun, but also to directly oppose
the downward pull of gravity. This is why most trees tend to be
straight, well-balanced, and, relatively speaking, tall. What is more,
they pursue these radical objectives tirelessly, in some cases for millennia. Viewed in this way, it could be argued that trees represent
aspiration and ambition in their purest forms. Simply by daring to

take root and grow, they bellow: "We refute gravity and entropy thus!"

Lots of people take inspiration from trees and forests, and we often like to think of them as sanctuaries of peace and tranquility. But this is deceptive; forests are, in fact, ruthlessly competitive places, where trees—and even branches on the same trunk—are engaged in life-and-death struggles for optimal position. The winner in this slow-motion race for space and light is determined by the tree or branch that photosynthesizes fastest and best. Photosynthesis, the process of manufacturing usable energy (carbohydrates) from sunlight and carbon dioxide, takes place in a tree's leaves or needles, and is an enormously complex process. Part of it involves the breaking down of carbon dioxide molecules. Our lives literally depend on this, because it is from this gas that the tree derives the oxygen we breathe. The overpowering need for sunlight is one reason West Coast conifers get so tall so quickly. Conversely, when such a tree is grown in isolation from its neighbors, it will concentrate on girth rather than height, resulting in a fatter, bushier version of its lean, competitive counterparts in the forest.

But no matter how vigorous a tree may look on the outside, this, too, is mostly illusion: like the earth's crust, the live portion of a tree is only a thin veil covering an otherwise lifeless mass. As counterintuitive as it may seem, a dead tree, shot through with molds, fungi, invertebrates, and bacteria, contains far more living material. The live portion of a healthy tree represents only about 5 percent of the total; the rest is just scaffolding, not unlike a coral reef. Beneath its leaves or needles, a tree is really a series of concentric tubes, each of which performs a specific function—defensive, vascular, or structural.

The outermost "tube"—the bark—functions much like our own skin: it protects the tree from external attacks such as animals, insects, and fire, and also helps to contain the fluids that keep the tree alive. Its thickness varies according to the needs of the tree; while the bark of a beech tree, for example, is less than half an inch thick, the bark on a big Douglas fir might be eight inches through. Douglas fir

thrives in the drier ground of the northwest where thick bark is helpful as a fire retardant. It is also heavy; loggers have been killed on
occasion by falling "walls" of this tree's bark. Just inside the bark is
the tree's vascular system, which is not much thicker than a piece of
cardboard. While photosynthate, which originates at the leaves or
needles, is drawn inward to feed the rest of the tree, additional nutrients are drawn upward from the earth in a matrix of water and distributed throughout the trunk and branches through a process called
transpiration. In this capacity, a tree operates like a giant straw with
many subdividing lines. In the case of a big West Coast tree, an individual molecule of water may take a week or more to travel from root
to branch, and yet such a tree can release hundreds of gallons of
water into the air each day. Under the right conditions, a forest can
generate its own fog and rain

Sandwiched within the vascular system is the cambium; only one
cell thick, it is this gossamer-thin "tube" of tissue that actually generates a tree's wood in the form of annual rings. Inside the cambium
and vascular layers is the "dead" central core of the tree; its cells may
hold and transport water, but they are not alive in the sense of being
actively engaged in the construction or maintenance of the tree. Over
time, the water in these cells is replaced with a rigid, epoxy-like substance called lignin, which gives a tree its strength. And this—the
tree's cellulose—is where the money is; an amazing variety of things
can be made from it—things as crude as charcoal and lumber, and as
refined as rayon and cellophane. Even so, when you compare the ele-
gance, economy, and complexity involved in the making of a tree
with our various attempts to exploit it, we look like so many cavemen
banging sticks together.

Photosynthesis is a true natural alchemy; it is what allows a plant
to, literally, build itself from air, water, and light—from "nothing."
This is an awesome feat on any scale, but it beggars comprehension
when one considers the sheer mass of material that must be generated
in order to "build" a sequoia, a redwood, or a Sitka spruce tree. In the
case of the golden spruce, however, the ability to do this was severely

compromised because any needles exposed to sunlight had their chlorophyll drastically depleted. Chlorophyll is the green pigment in leaves and needles and it is what makes photosynthesis possible. In terms of its ability to convert energy, the golden spruce's impairment could be compared to a person with lungs that function at a third of their normal capacity. For this reason, no one is quite sure why the golden spruce was able to compete so well against healthy trees for three hundred years, or why it was able to grow to over 160 feet tall.

A tree that exhibits this pronounced yellowing is called a chlorotic, and while it is not uncommon to see a chlorotic branch— or "sport"—on an otherwise healthy specimen, it is impossible, in the theoretical sense that the flight of bumblebees is impossible, for an entire tree to be chlorotic and survive. Chlorosis is directly related to the health and well-being of the carotenoids—hydrocarbons which form the red, yellow, and orange pigments found in all photo-synthesizing cells. As unfamiliar as their name may be, most people can recognize them at a glance; it is carotenoids (from the same root as "carrot") that are responsible for the brilliant fall colors in decid-uous (leaf-bearing) trees. While these pigments are present through-out the year, they only become visible as the leaves die back in win-ter because they break down more slowly than the green chlorophylls that ordinarily dominate a leaf's color. In conifers, however, they play a more modest, supporting role; under normal circumstances this order of tree species seldom reveals its carotenoids in any obvious way—hence the nickname "evergreen." Exceptions to this rule occur most often in the cases of death and disease.

Chlorosis can be caused by any number of things, including infer-tile soil, bug infestations, girdling (the typically fatal removal of a strip of bark all the way around the tree), and by too much, or too lit-tle, sun and/or water. But the golden spruce suffered from none of these afflictions. Not only was it big and old, which translates to "suc-cessful" in Sitka spruce terms, but it was growing in prime spruce habitat under ideal conditions. All the trees around it were healthy. Lacking external causes for chlorosis, all the evidence points to the condition originating within the tree itself. Rather than suffering from some pathology, the golden spruce probably had some inherent flaw, most likely one that affected the carotenoids. One of the several functions served by carotenoids is that of a barrier to ultraviolet light; in this capacity they act as a kind of natural sunscreen—a localized ozone layer—to protect the more UV-sensitive chlorophyll. In a plant where the carotenoids are not blocking UV rays as they should, the chlorophyll will break down and the plant will die. As long as such a tree remains shaded, its defective carotenoids won't be tested, but as soon as they are exposed to direct sunshine, the flaw is revealed. As the undefended chlorophyll deteriorates, the green in the needles is lost, leaving only the faulty yellow carotenoids, which are unable to photosynthesize on their own. Under ordinary light conditions these yellow needles (which are still alive) will usually burn out and fall off. Chlorosis of the kind exhibited by the golden spruce bears some similarities to albinism, but a closer analogy can be found in xeroderma pigmentosum, the exceedingly rare skin disease that makes ultraviolet light fatal to humans. Though profoundly disruptive to a normal life, a person with this disease can protect himself by avoiding sunlight. However, a tree with this condition finds itself in a lethal catch-22: in its instinctive quest for light it grows itself to death.

Somehow the golden spruce defied all logic by growing tall enough to be exposed to the sun's full force without being killed in the process. Nor was it seriously stunted or delayed in any way; it was the same size as a normal tree of its age would have been under those growing conditions. And coloring wasn't the only way in which the tree distinguished itself; as the golden spruce grew to maturity it revealed another peculiarity. Normal Sitka spruce trees are not only promiscuous—they will interbreed with any other spruce that will have them—they are also hermaphroditic, meaning each individual produces its own ovules as well as the pollen to fertilize them. But the golden spruce produced neither, making it, in effect, an asexual, infertile one-off; the chances of such an accident occurring again—successfully—are almost incalculably small.

Not only was the golden spruce sterile and a radically different color than its normal counterparts, it also assumed a markedly different shape. As noted earlier, Sitka spruce are not particularly tidy trees; unlike many other conifer species, their natural tendency is toward frowsy asymmetry. The golden spruce, however, possessed a hedgelike density and an uncharacteristically conical shape. "It was perfect," recalled a Haida silversmith and tree faller named Tom Greene. "It looked like a manicured tree." An American forester and soil scientist named Edmond Packee who spent several years in Haida Gwaii and was familiar with the tree speculated that its compact, tapered appearance was a spontaneous adaptation intended to minimize the UV exposure that more typically wayward branches have to endure. Supporting this theory are photos of the golden spruce that show the dead and bleached remains of limbs that tried to venture out beyond the golden safety zone.

Plant physiologists, like doctors, have difficulty explaining strange behavior in the absence of pathology. How does one make sense of a specimen that looks "sick" but isn't? The only way botanists know to explain an oddity like this is with the term "mutation." But, without analyzing a plant's DNA, this is a vague explanation at best. A tree, like a person, can in theory be riddled with mutations that are invisible; as long as they don't impact the individual's appearance or health, they can go undetected. In an effort to get to the bottom of the mystery, a young forester named Grant Scott wrote his undergraduate thesis on the golden spruce. While he was enrolled in UBC's school of forestry in the mid-sixties, he spent two summers cruising timber in Haida Gwaii. During that time he got to know the Yakoun Valley very well; not only did he acquaint himself with the golden spruce, but he found its biggest and best-known counterpart about ten miles upriver in a nearly identical position on the east bank. This Sitka spruce, while technically golden, is less uniformly colored and more typically spruce-shaped; about 100 feet tall and as many years old, it is, one could say, more like what one would expect of a mutant under those circumstances. There are, in addition to this one, a couple of other "golden" spruces rumored to live on the islands, but like the one Grant Scott found, none of them are as big, yellow, or uniquely shaped as the legendary tree that grew at the north end of the Yakoun.

Back in Vancouver, casting about for a thesis topic, Scott realized that what he really wanted to investigate was this extraordinary tree. But this would be a hard project to sell to his professors, most of whom were focused on the logging industry. Oscar Sziklai might have been a possible adviser if he hadn't been too busy to trouble himself with undergraduates. It was at this point that a young professor from Yale named John Worrall entered the picture. Worrall was another Englishman, a contemporary of Bruce Macdonald's, who had been hired by UBC to teach plant physiology. He actively encouraged Scott to pursue the mystery of the golden spruce and agreed to sponsor him. The goal of Scott's thesis was to determine (a) why the golden spruce was golden, and (b) how the tree could survive with such a crippling disability. According to Scott's research, it all comes down to chloroplasts, the tiny, subcellular bodies that do for plants what photovoltaic cells do for solar-powered machinery. It is the disk-shaped chloroplasts that generate chlorophyll, thus enabling photosynthesis to take place. The needles on a tree are essentially vehicles for the chloroplasts, and they function much like solar panels; they are so well designed that chloroplasts will actually reorient themselves within a tree's needles throughout the day in order to take full advantage of the sun. The flaw, believed Scott, lay in the proteins that bond the chloroplasts together. They functioned normally until they were exposed to sunlight, at which point they would mutate, causing the chloroplasts to become dangerously inefficient. Fortunately for the golden spruce, the needles that weren't directly exposed kept their integrity and were able to live—even thrive—on reflected light.

Grant Scott's experiences with the Haida and the natural force of the islands made such an impact on him that instead of pursuing a career in the logging industry, he has become a negotiator and forestry adviser who works exclusively with northern coastal tribes.

"Every time I go back there, I feel just the way I did the first time," he explained from his home on a small island in Georgia Strait. "You just want to go and see what's over that next hill. Of course," he added, "you know what's over there now." He was referring to clear-cuts.

Trees, like people, mutate all the time, but with each roll of the chromosomal dice there are heavy casualties. The mutation that likely caused the golden spruce isn't all that uncommon—most tree breeders have encountered golden seedlings at one time or another— it is the Yakoun specimen's vigorous survival that is truly freakish. And it points to such a mutation being uniquely adapted to the heavy cloud cover of Haida Gwaii, which are also known as the Misty Isles. Its location next to the river may have met other preconditions for survival as well: in addition to exceptionally fertile floodplain soil, it could have benefited from a phenomenon called albedo.

When solar radiation encounters an object, it is either absorbed or reflected—usually a bit of both; the percentage that is reflected is called the albedo, and it fluctuates according to the reflectivity of the substance in question. Fresh snow, for example, has an albedo of 75 to 95 percent, which is why alpinists can get sunburned on the insides of their nostrils; a highway, on the other hand, has an albedo of 10 to 15 percent: it doesn't reflect much, but it gets hotter than hell—just like beach sand. Sunlight reflected off of water still contains every-thing needed to facilitate photosynthesis (the visible portion of the light spectrum is called photosynthetically active radiation, or PAR), but its albedo fluctuates depending on the sun's angle: low-angle morning and winter light has a much higher albedo—as much as 100 percent—while the midsummer noonday sun has an albedo of less than 10 percent. The state of the water's surface is also a factor, but the Yakoun is glassy smooth where it flows past the tree so it wouldn't have reduced the potential albedo in an appreciable way. Given that there are lots of other tall trees in the immediate vicinity, the only sunlight that finds its way to the surface of the Yakoun is higher-angled midday and summer sun, which translates to a low albedo—just what the doctor might have ordered for a UV-intolerant tree like the golden spruce. It is conceivable that even though the golden, skyward-facing needles were out of commission, the green needles underneath may have been nourished by the albedo bouncing up from below. And though they were dysfunctional, the golden needles may have contributed something as well; while the albedo of a typical conifer forest is only around 10 percent, the golden spruce's was much higher. Its needles were so reflective that in video footage of the fallen tree shot with a camera light, they are actually blinding. Perhaps, then, the golden spruce's defect also helped to keep it alive, by reflecting a higher—but not lethal—percentage of the albedo onto the unaffected needles.

But even if this turned out to be the case, so what? From the point of view of the Haida's oral amalgam of history, myth, and parable, such speculation amounts to little more than a parlor game for botanists. If one were to analyze the mathematical chances of a single tree having not one but at least three highly visible defects that impacted not only its physical structure and ability to photosynthesize but its ability to reproduce as well—and then factor in the likelihood of their occurring in an environment that would somehow enable the tree to survive and flourish in spite of them—one would come up with odds bordering on the infinite. The word "miraculous" might legitimately come to mind, and in a version of the golden spruce story that probably predates the one about the disobedient grandson, a miracle is exactly what was requested.

Hazel Simeon is a Haida artist and a maker of button blankets—ceremonial cloaks—which she sells to the Haida and also to collectors of Haida art, and her specialty is blankets depicting the story of the golden spruce. She speaks Haida fluently and was one of the last islanders to be raised in anything resembling a traditional fashion. As a child in the 1950s, she was told by the elders in her family that she was to be a blanket maker. "They wouldn't even let me cook or fish," she said. "They didn't want me to be distracted."

Traditionally, button blanket makers are women, but the tight, formulaic designs they use are almost always drawn by men; they are then appliquéd, typically, with black and red felt outlined by plastic or abalone buttons. But Simeon's blankets are altogether different; rendered in wool, cotton, buckskin, and suede, they may be decorated with golden beads, disks of copper and brass, buttons of abalone, stone, bone, and anything else she is given or manages to find. When she makes a golden spruce blanket, its trunk is also the torso of a man or a woman, depending on which part of the golden spruce story she is telling. According to Simeon, the first tree was a woman and the second tree—the one Hadwin cut down—was a man: the woman's nephew. They were the sole survivors of a smallpox epidemic; it was clear to both of them that their clan was doomed and that the magic, as Simeon puts it, was finished. Because of this, they requested that the spirits leave a sign that this magic had once existed so that future generations, whoever they might be, would understand who had lived there and the power they had once wielded. The aunt died first and the nephew buried her on the banks of the Yakoun. A golden spruce grew over her grave, and it was a "female"; it grew for "about three hundred years" before being struck by lightning. The nephew, by this time, was very old and not feeling well, so he went to his aunt's grave to wait for death to come. When he died, a second golden spruce grew. This was the sterile male—the last golden spruce.

Time and events are clearly elastic in this version of the story, but it is still tempting to ask if such miracles could occur; they certainly do in the Bible, a text with which most Haida storytellers of the past century are familiar. Still, common sense would say not, and yet scientists have demonstrated that under ideal conditions, a spruce scion can take root simply by being plunged into receptive soil. It would be hard to find soil more receptive than that of the Yakoun Valley. Eagles and ravens are common here, often perching in the tops of trees, where they will clip branches with their powerful beaks. It is conceivable, then, that a clipped or broken twig from the top of the "first" golden spruce could have found its way, stem foremost, into the rich humus of a rotting nurse log or the forest floor. The chances are minute, of course, but no more so than the chances that the golden spruce would have grown in the first place. It is exactly this willingness to host the implausible that makes the islands and their surroundings so extraordinary.

Until a hundred years ago, the golden spruce coexisted with the only caribou known to have lived in a rainforest environment. Before 1908, when the last four specimens were shot by hunters, Graham Island was home to Dawson's caribou, a subspecies that was probably stranded on the islands following the last ice age. On the other side of Hecate Strait, in a confined area around Princess Royal Island and the adjacent mainland, lives a unique population of white black bears. According to scientists, these kermode bears are the result of a recessive gene—not albinism; they make up about 10 percent of the local bear population, and they interbreed normally with their black counterparts.

Nearby, in the same waters that support the world's largest octopus (the Pacific Giant), are the last known vestiges of the most massive entity that ever lived. The first signs that something huge and remarkable might be living in Hecate Strait showed up in 1984 while scientists with the Geological Survey of Canada were doing a seafloor mapping exercise. Using sonar imaging, they observed certain acoustic anomalies that generated what was described as an "amorphous, irregular seismic signature having no coherent internal reflectors." The source of the cryptic message turned out to be a vast prehistoric sponge that covers hundreds of square miles of sea bottom in Hecate Strait, southward to Queen Charlotte Sound. Before this remnant was discovered, silicious ("glass") sponge reefs of this kind were believed to have been extinct for 65 million years. During their most successful era in the late Jurassic period, 140 million years ago, they covered hundreds of thousands of square miles of what was then the ocean floor; their fossilized remains have been found from Romania to Oklahoma. Meanwhile, 150 miles southwest of these sponge reefs and more than a mile beneath the surface, an isolated group of volcanically heated vents is generating the highest liquid water temperatures ever measured in nature (over 700 degrees Fahrenheit). Surrounded by a virtually lifeless deep-sea desert, these thermal "oases" support a bizarre ecosystem teeming with hundreds of thousands of creatures per square meter.

IN 1977 GORDON BENTHAM managed to get hold of some golden spruce scions from a Vancouver Island nurseryman; like Sziklai's, they had been taken from about halfway up the tree and exhibited the branchlike plagiotropy common to the species. Once again, only a tiny percentage of the cuttings survived (in private gardens), and among them was the pair that Bentham gave to Roy Taylor in 1983. At the time, Taylor tucked the five-year-old trees into a shady, out-of-the-way spot in UBC's botanical garden and hoped for the best. A decade later the trees were still alive, but they were only about 6 feet tall (an ordinary spruce would be close to 50 feet tall by this time). It was at this point that a UBC gardener named Al Rose took it upon himself to move the golden dwarfs to a somewhat sunnier location, and it was there, next to a quiet path in the garden's native plant section, that Taylor's successor, Bruce Macdonald, found them. Within twenty-four hours of finishing the *Sun* article about the felling of the tree Macdonald was taking calls from CNN, the *New York Times*, and film crews from as far away as Germany and Japan.

Macdonald immediately notified the Haida tribal council and offered them one of the trees, but this raised a host of questions that neither he nor the Haida had easy answers to. First of all, the cuttings had been taken without Haida permission so they were, in the eyes of some, stolen property; what, then, was the appropriate response to an offer of their return in a radically altered state? Second, the trees had been grown off-island; if they weren't nurtured by the Yakoun, were they really the same golden spruce? These questions came at a time when North American tribes had begun challenging the rights of museums to the bones and artifacts that were exhibited in their halls and stored in their basements. The Haida, in particular, have had good success repatriating some of this material, but they were ambivalent about Macdonald's well-intentioned offer. However, they showed enough interest that Macdonald ordered the healthier of the two specimens to be dug up and prepared for shipment to the islands. With its root ball wrapped in burlap, the tree was then taken to the garden nursery, where it was set in a bed of sawdust and left alone except for regular watering. Meanwhile, Canadian Airlines was contacted and they offered to airlift the tree to the islands free of charge. Everything was in place down south, but there was still disagreement within the tribal council about where the tree should go and who should administer it. As the debate wore on, the storm around Hadwin and the tree blew itself out; other issues, among them the continued clear-cutting of the northern islands, pushed their way to the fore. Ordinarily a six-foot Sitka spruce could survive "in storage" almost indefinitely, as long as it was watered, but UBC's golden spruce was much less stable. Badly stressed by the move, it began shedding its needles, and within six months the tree was dead.

While Macdonald had been making plans to move the UBC specimen, the Haida had been consulting with local foresters at MacMillan Bloedel, who were supporting the Haida in their efforts to save the tree, and plans were being made to take new cuttings. As it turned out, this was the only upside to an otherwise tragic situation: if Hadwin had cut the tree at any other time of year, there would have been virtually no hope of saving it. Scions can only be taken between the months of December and February, when the trees of the North Coast are dormant. The winter months also mark the critical threshold between the summer formation of the next year's buds and the springtime when they flush, or sprout. The bud is the key to a scion's success; without one, a scion or, for that matter, a tree has no motive to carry on. Once again, the fate of the golden spruce resided in a tiny, kinetic bundle waiting for a highly unlikely set of circumstances to set it in motion, just as it had three hundred years before.

Another advantage to the tree being on the ground was that it would now be possible to take the most promising scions from the very top where apical dominance—the upward growing impulse—is strongest. While there was some debate among the Haida leadership about whether to try to revive the tree or simply let nature take its course, there was no time to lose and those in favor of taking cuttings prevailed. Even so, it could still be a moot point: given the previous record of propagation attempts, there was no guarantee that these would fare any better. Erin Badesso, a forester for MacMillan Bloedel who was based in the islands, made arrangements with the Ministry of Forests' Cowichan Lake research station at the south end of Vancouver Island; he then took about eighty cuttings from the tree as it lay dying on the bank of the Yakoun. Amputated tree limbs are treated much like those of human beings: after being wrapped in wet newspaper and plastic bags, the golden spruce cuttings were packed in ice-filled coolers and flown south where they were divided up between three different propagators who would use a variety of methods in order to maximize the odds of success. The bulk of the cuttings were given to Luanne Palmer, an expert grafter at the Cowichan Lake research station, who dropped everything and set to work. Palmer had the sense that she was engaged in a unique under-taking when she unwrapped the scions, which, even in their semi-frozen state, retained their striking golden qualities. She had done grafts like these thousands of times before—as many as six hundred in a single day—but never had the stakes been so high. When Sziklai took his cuttings, the parent tree had been alive and well; this time, if the grafts failed, there would never be another chance.

Grafting is an ancient and surprisingly simple process: as one gar-dener put it, "All you do is attach two wounds together." However, it helps to have a green thumb and an accommodating plant; roses and fruit trees are the most common candidates, but many conifers are also receptive. Palmer was going to use a side veneer graft, a method that involves attaching a scion about two inches long to the stem of a normal Sitka spruce seedling. In the case of a side graft, both scion and rootstock have their bark cut away at the point of contact and are then bound together with ordinary rubber bands; afterward a drop of wax is daubed on the high side of the joint to keep excess water out. Palmer did this about forty times while another forty cuttings were set directly into specially prepared soil. Then, as all those with a stake in the golden spruce held their breath, the little clones were taken to the greenhouse, where nature would take its course. Assuming the scions survived the grafting and planting processes, it would be at least two months before they flushed, indicating that the scion was viable and growing. Once over this hurdle, it would still take at least six months of careful watering and fertilizing before it would be deemed safe to start pruning back the rootstock's other branches in order to encourage the golden scion to take over as the leader. Even if it survived this step, it would be an additional two or three years before this somewhat Frankenstein-like golden spruce was ready for transplanting. During this lengthy and labor intensive process, there was a lot of time and space for things to go wrong, but no one doubted it was worth the risk or the trouble. What these cuttings promised that Sziklai's and Bentham's hadn't was a scion that would grow like a tree rather than a branch. If Palmer's grafts took and things went well, a true golden spruce might once again grace the Yakoun.

CHAPTER THIRTEEN

Coyote

... beasts
Got the buddha-nature
All but
coyote.

—Gary Snyder, "How Rare to Be Born a Human Being!"

IT WAS IN APRIL, just as the search for Hadwin was slowing down, that the buds on Luanne Palmer's golden spruce scions flushed. And it was in mid-June, when an inch of new growth had been added, that his kayak and camping gear were discovered on Mary Island. To those on the outside, it might have seemed, then, that the golden spruce's chances of survival were a lot better than Hadwin's. But rather than confirming his death and closing the case, the discovery on Mary Island rekindled old suspicions, both in the minds of police and those who knew Hadwin personally. While the debris might have been evidence of a legitimate misadventure at sea, it seemed equally plausible, based on what the investigators now knew about this man, that the wreck had been staged. If one stopped to think about it, there were any number of ways a kayak could have found its way onto the rocks at Edge Point.

According to computer-generated scenarios, Hadwin's kayak could have drifted into Revillagigedo Channel from almost anywhere in Hecate Strait due to the season's prevailing winds. This raises a host of possibilities, ranging from Hadwin's having capsized en route to Masset, to Constable Walkinshaw's suggestion that he "could have got a pumping" (been shot) out on the water. Given the amount of shipping traffic in that area, he also could have been struck by another vessel—by accident, or intentionally. Whatever happened, it is a near certainty that Hadwin's kayak remained intact until it hit the beach. With the cockpit full of water, it would have been over-loaded, and the combination of logs, boulders, and wave action would have broken it up quickly, releasing its contents, most of which were found nearby. Missing were Hadwin's paddles and pump and, of course, Hadwin himself. Not surprisingly, he wasn't wearing his life jacket and probably never had, as it was found on Edge Point in unused condition. Hadwin's food was nowhere to be found, but animals could have scavenged that within hours.

In the event of a capsize, Hadwin wouldn't have known how to roll his kayak back upright while staying in the cockpit, a tricky maneuver at the best of times, particularly in heavy seas with a loaded eighteen-foot boat. This means, he would have had to make what kayakers call a "wet exit," after which he would have had to turn the boat over manually and then climb back in. The front and rear hatches were watertight so the boat would have floated, but the cockpit would have been awash. Even if he had managed to pump out the cockpit successfully (a challenging procedure in rough conditions), the hypothermic clock—already ticking—would have been in imminent danger of running out. Body heat is lost twenty-five times more quickly in water than in dry air, and the average water temperature in Hecate Strait, in February, is around 40 degrees Fahrenheit.

Even without factoring in windchill, this would give an average person about half an hour before he became incapacitated, and another hour or two before he lost consciousness altogether. Hadwin, with his high tolerance for cold water, rigorous discipline, and good conditioning, might have lasted far longer, but unless he was very close to shore, this would have only prolonged his suffering. If he did roll, the combination of waves, fog, and/or darkness could have left him completely disoriented, and even if he had land in sight, any combination of contrary winds, waves, or current could have swept him away with ease.

In the event that Hadwin actually made it ashore, he would still need a heat source in order to stave off full-blown hypothermia, and based on his bare-bones inventory (dry matches and coffee) following his Kruzof Island sojourn, this is something he may well have had. Despite his risky behavior and apparent imperviousness to cold, Hadwin had thirty years of solo wilderness experience under his belt; he would have been keenly aware of the dangers of hypothermia. Had he managed to stabilize his body temperature, he would have been good to go, perhaps indefinitely. His old housemate and colleague Paul Bernier was impressed by more than Hadwin's stamina in the woods. "I know a lot about him," he said, "and I know he could survive on just about next to nothing. He knew what kinds of plants were around; he introduced me to different plants that we could eat."

Cory Delves, one of Hadwin's former supervisors at Evans Wood Products, agreed. "Basically," he said, "you're dealing with a person who, with very few resources, could be dropped anywhere on earth and come up smelling like a rose."

The North Coast isn't just a good place to hide dead bodies, as trooper McPherron observed, it can absorb live ones too, and there are more than a few who believed it absorbed Hadwin. This may have been what he ultimately wanted: total immersion in the environment where he felt most at home, and most himself. If so, it aligns him squarely with many of the Haida, who, under different circumstances, might have been his allies and even his protectors. "There's no planning," said Guujaw, the current president of the Haida

Nation, to another kayaker when they were discussing local logging practices. "One resource after another, they create a license and wipe it out. It's not a commodity; there is no commodity. Just leave your kayak, take off your shoes, and go in there."

And this is exactly what Hadwin might have done. Somewhere past Port Simpson, he may have landed on the wild, empty coast, given his boat a push off the beach, and walked into the forest.

Supporting this theory, in the minds of Sergeant McPherron and his Canadian counterpart, Corporal Gary Stroeder, was the state of the wreckage, which did not seem consistent with the time lag, or the rugged environment in which it was found. This complicates matters considerably, as do two other details: first, that under typical winter drift conditions, Hadwin's empty kayak should have washed ashore within a matter of days, and, second, that Hadwin charged three hundred dollars' worth of food on the day of his first departure—an awful lot of provisions for what was supposed to be a five-day journey. But there was something else that Corporal Stroeder found even more troubling: the location of Hadwin's ax. How, he wondered, did such a heavy object get above the high-tide line?

Sergeant McPherron was puzzled by this, too, and it led him to speculate that Hadwin might have broken up the kayak himself, in order to make it look like an accident. While the large food bill and the relative lack of abrasion on the boat's hull support this theory, one wonders why Hadwin would have gone to these lengths on an island that was a five-mile swim from any other significant landmass—unless he intended to live there for a while. Mary Island covers eight square miles and has never been logged; food and fresh water are abundant, and no one has ever searched it with an eye toward finding someone who wanted to remain hidden. But if that's not what happened, then what's the alternative—that a bear carried the ax up the beach in its jaws? Could the castaway Dennis Harrington/Roe have moved it? Not likely, because he was rescued from the opposite side of the island, and given his footsore condition, he wouldn't have traveled far. There is little doubt that Walker was the first one to happen upon it, and he found it only by chance; even after he gave the Coast Guard specific directions to the site, they were unable to locate it until he went back and nailed large pieces of the kayak to a tree. Among Hadwin's effects was a shaving kit that contained a bottle of medication. According to Sergeant McPherron, the label was illegible save for Hadwin's name, so they threw the bottle away. He didn't remember if it was empty or full.

CORA GRAY, who has good reason to take stock in dreams, woke up one morning having seen a man in a green raincoat floating facedown somewhere off the coast. But it's hard to know how literally to take these found images; She may indeed have seen someone, perhaps the same person for whom Hadwin's dental records were requested, but Hadwin's raincoat was yellow, and he wasn't wearing it at the time he parted company with his kayak; we know this because it was found on Edge Point along with the rest of his rain gear. Nonetheless, such a fate is a plausible, even likely one under the circumstances, but not according to another clairvoyant who believes she saw Hadwin too. Hadwin's wife, Margaret, devout Christian though she is, was moved to consult a psychic on two separate occasions. The seer saw Hadwin alive, in southern B.C.; he was in poor health, she said, working only for food. Somehow this doesn't sound like him, even under duress.

Of the various alleged Hadwin sightings made up and down the coast during this period, one in particular excited the authorities on both sides of the border. On August 31, more than two months after his kayak was found, a man matching Hadwin's description was spotted boarding a ferry in Pelican, Alaska, a tiny fishing community on Chicagof Island, just north of Sitka. Apparently he had been kicked out of town. The call to the Prince Rupert RCMP came following *six* separate confirmations that this individual matched Hadwin's missing person's photo, which had been posted in town. The ferry was bound for Juneau, the state capital, and a Sergeant Tyler met the suspect there; when questioned, he claimed he was an archaeologist from Prince Rupert, and was on vacation. Sergeant Tyler was appropriately skeptical and he took prints of both the man's thumbs, which he faxed down to Prince Rupert. The thumbprints, too, were uncannily similar to Hadwin's, but in the end it was determined that they belonged to a different man. Since then, neither Hadwin nor anyone resembling him has attracted the attention of the authorities, so rumors have filled the void:

He was killed by Indians.

He's running a trapline outside Meziadin Junction (a wilderness crossroads east of Hyder).

He was seen on Wrangell Island (between Sitka and Ketchikan).

He's in jail in the States.

He's in Siberia.

Hadwin's younger children—now in their twenties—clung to the hope that their father was alive for years. They are victims of what psychologists call "ambiguous loss"; it is devastating to lose a parent under any circumstances, but to not know with any certainty if he is truly gone, or if he might one day come home, is particularly cruel and painful. Margaret, on the other hand, has been trying to have her husband declared dead. Constable Walkinshaw believes that Hadwin could be alive: "The whole cop in me is saying there's something too neat about this." Most of the Haida feel the same way, and so do a striking number of people who have known Hadwin personally over the years: "He could be anywhere," speculated Corey Delves, "from the Fraser Valley to Prudhoe Bay."

"We all think he's alive," said Al Wanderer, "anyone who had anything to do with him. He's a survivor."

These aren't idle claims; one of Hadwin's previous bosses was reluctant to discuss him, even years after the fact, because he feared that Hadwin—wherever he was—might find out and come after him if he said anything critical. Constable Jeffrey firmly believes Hadwin drowned, but Corporal Stroeder isn't so sure. "If a coroner asked me to justify that he was dead, I wouldn't be able to," he said. "There are too many loose ends."

One of them stretches all the way to California. Sometime during the Thanksgiving weekend of 2000, someone made a nearly fatal chain saw cut in Luna, the massive Humboldt County redwood made famous by the environmental activist Julia Butterfly Hill, who spent two years living in the tree's branches. As with the golden spruce, the cut did not fell the tree, but it left it extremely vulnerable to high winds (it has since been reinforced with heavy steel brackets and is still alive today). The cutter was never caught. However, as similar to Hadwin's modus operandi as it seems, the attack on Luna was almost certainly the work of a local logger who was enraged, not by Charles Hurwitz, the rapacious absentee landlord who was liquidating that piece of forest in order to pay off other debts, but by the self-righteous meddling of environmentalists in what most West Coast loggers feel is a God-given right. "Eight hundred years to grow, and twenty-five minutes to put on the ground," as one veteran B.C. logger put it. "It's sad, but it's a living."

DURING THE LAST YEAR before his disappearance, Hadwin began referring to himself, on occasion, as Coyote; sometimes he even signed his letters this way. In the end, he may have had more insight into himself than most people gave him credit for. He certainly would have known these creatures well, and he wasn't the first to see the similarity. Hadwin had that same fast, tireless, unkillable quality that coyotes seem to have. Unlike wolves, coyotes will hover on the outskirts of civilization, darting into inhabited areas as need arises before ghosting away again into the bush. If one had to sum up Hadwin's life in one sentence, it would look much like that. Hadwin's case is still considered open by the Alaska state troopers and the

Mounties (such files have a shelf life of twenty years), but no one has bothered to search for him in Siberia. Cora Gray recalled that Hadwin "talked about Russia a lot. He'd say, 'If I was going to choose a place to stay, it would be in Russia. Don't be surprised if you hear from me from there.' So now, when the phone rings late at night, I don't answer."

CHAPTER FOURTEEN

Over the Horizon

A culture is no better than its woods.
—W. H. Auden, "Bucolics II: Woods"

AL WANDERER, HADWIN'S former colleague from Lillooet, could have been speaking for all woodsman throughout history when he looked back over his own empty corner of British Columbia and said, "Good God, I didn't think it was possible to log this much." Anyone who has traveled in the woods of the Pacific Northwest would know exactly what he meant. Even now these forests have an infinite feel —until you see the clear-cuts and realize how extraordinarily efficient humans can be at altering the landscape. Out here, the empty spaces still look like wounds, like violations of the natural order, but back east—that is, from Chicago to Babylon—we find this hard to visualize because the clear-cutting happened generations before any of us was born. Treeless expanses look normal to us—"natural," even. We tend to look at time in a myopic, human-centered way, but trees offer an alternative means of measuring our progress (as well as our regression). Growing at a rate somewhere between stalagmites and human beings, forests can serve as a kind of long-term memory bank, revealing things about our environment, and even ourselves, that only our great-great-grandparents could have told us. The short version of the forest's message was well paraphrased by historian John Perlin: "Civilization has never recognized limits to its needs."

In fact, the realization that the New World was not a bottomless cornucopia intruded surprisingly early on. By the 1630s, the beaver had already been extirpated from much of the New England coast, forcing fur traders to probe westward and northward, ever deeper into the forest. In 1640 the first deer-hunting ban was enacted (in Rhode Island) in an effort to preserve the plummeting deer population. Thickly settled areas like Boston and southern Manhattan were being forced to import firewood from elsewhere on the coast well before 1700. At this time, a typical fireplace sent about 80 percent of its heat up the chimney and might consume twenty cords of wood per year (about a month's work of cutting, splitting, and stacking for one man). William Strickland, who could be described as a forefather of the commodities analyst, was an early critic of the prevailing attitude toward trees: "What is not wanted for any present purpose is set fire to," he observed at the end of the eighteenth century; "if care not be taken it will soon be very scarce. . . ." But such prescient finger-wagging had little effect; the notion that North American timber might be finite seemed laughable—that is, until 1864, when a ground-breaking book called *Man and Nature; or, Physical Geography as Modified by Human Action* was published by George Perkins Marsh. Marsh was a Renaissance man from Vermont, and he has been called America's first environmentalist. In *Man and Nature* he laid out, in no uncertain terms, the negative impact of human behavior on the natural landscape: "Man," he wrote more than 140 years ago, "who even now finds scarce breathing room on this vast globe, cannot retire from the Old World to some yet undiscovered continent, and wait for the slow action of such causes to replace . . . the Eden he has wasted." That year (1864) saw the creation of California's Yosemite State Reserve, which included the continent's first federally protected trees.

There was ample evidence to support Marsh's thesis. With the westward push fully under way, the great oak and pine forests of the Lake states and southern Ontario were melting away before a no-holds-barred assault of fire and steel. Within a decade government and scientific bodies were sounding regular alarms, warning any who would listen about the dangers of timber waste, fire, and soil erosion that dogged logging and land-clearing operations the way vultures and coyotes followed the buffalo skinners. Those working closest to the land, in the nascent sciences of geology and forestry, were horrified by what they saw. "Nearly the entire territory has been logged over," wrote one forester in 1898, describing the woods of northern Wisconsin.

> The pine has disappeared from most of the mixed forests and the greater portion of pineries proper has been cut. . . . Nearly half of this territory has been burned over at least once, about three million acres are without any forest cover whatever, and several million more are but partly covered by the dead and dying remnants of the former forest.

Further west, on the Great Plains, the buffalo population was meeting the same fate: by the 1880s, the most numerous herd species on earth— once numbering in the tens of millions—had been reduced to fewer than 300 individuals. It was as if the New World had been invaded by legions of sorcerer's apprentices: while they were able to summon up the world-changing energies of the steam engine, the circular saw, and the Sharps .50 caliber rifle, they failed— or simply refused— to grasp the greater implications of such super-human capability.

Europeans had been through this before, and though it had taken them centuries to accomplish what the North Americans were doing in decades, they had fared no better. Their own native bison had been wiped out long since (the current population of around 3,500 European bison was generated from just 5 survivors). As it turned out, many European forests were brought back the same way the bison were—by systematic breeding. Silvaculture, the science of farming trees, was born in England in the mid-seventeenth century; its principles were quickly adopted by Europe's scientific community, and by the mid-nineteenth century tree plantations were becoming widespread throughout the continent (stands of Douglas fir have been growing in Belgium since the 1880s). Silvaculture then made the jump across the Atlantic, but while "new" forests were soon sprouting up in city parks, and even on the treeless Plains, the science wasn't applied to the New World's stump fields until the 1920s, and then only in tentative, experimental applications.

In the early 1890s, while John Muir was founding the Sierra Club, "cut and run" logging communities were already turning to ghost towns in Idaho as their former residents pushed westward to the coast. By 1919, just as a group of wealthy Californians was forming the Save the Redwoods League, the first portable chain and circular saws began appearing on the cover of *Scientific American*. Six years later "lady conservationists" were actually tying themselves to doomed redwoods while huge machines like the Washington Flyer were hauling trees out of the Northwest forests as fast as chokermen could cable them up.

What the chain saw and its mechanical attendants—the bull-dozer, log skidder, and self-loading logging truck—have done is to reduce the great trees of the Northwest down to objects that a man of average size and physical condition can fall, buck, load, and trans-port. Today, a tree ten feet across the butt can be felled in ten minutes flat, and bucked up in half an hour. Afterward it is a matter of moments for a grapple yarder—essentially a huge mobile claw on caterpillar treads—to pick up the multiton logs and load them onto a waiting truck (no need for a spar tree anymore). In theory, then, a

200-ton tree that has stood, unseen, for a thousand years and with-
stood wind, fire, floods, and earthquakes can be brought to earth, rendered into logs, and bound for a sawmill in under an hour—by just
three men. In 1930 it would have taken a dozen men a day to accomplish the same thing. In 1890 it would have taken them weeks, and
in 1790 it would have been a matter of months—assuming they were
even able to fell the tree.

Meanwhile, smaller timber can be harvested by feller-bunchers—
logging's equivalent to the combine harvester. These frighteningly
efficient devices can drive through a forest, cutting, limbing, and stacking trees in a single continuous motion. When first introduced in the
1960s, they worked only on open, level ground—a type of terrain that
is in short supply on the Northwest Coast, but lately models capable of
handling three-foot-diameter logs on 30 percent grades have been
developed. Equipped with powerful headlights, they can operate
twenty-four hours a day. Safely belted in behind the joystick of such a
machine, a logger can now roll through a mountain wilderness in air-
conditioned, stereophonic comfort, harvesting the forest at a rate—and
at a remove—that his grandparents never would have dreamed of.

Even Bill Weber, who has only been working in the woods since
the late 1970s, expressed astonishment: "I never dreamed the old
growth would be finished," he said. Much of the wood he is cutting
today would have been scoffed at by his parents' generation. "Twenty
years ago, we'd have looked at the wood we're into now and say, 'What
the hell are we doing in this shit?'"

One of Weber's colleagues, Earl Einarson, a fifty-four-year-old
tree faller, expressed the logger's conundrum as honestly as anyone.
"I love this job," he explained, gesturing toward the wild chaos of the
old-growth forest he was in the process of leveling. "It's a challenge to walk into a mess like this and get it looking civilized."
(This child of the atomic age would have won a sympathetic nod
from any seventeenth-century settler.) Einarson paused for a moment and Weber, his supervisor, looked over his last falling cut while a
big glossy raven lighted on a nearby branch that would no longer be

there in another twenty-four hours. A hundred yards away, an unknown and unnamed waterfall tumbled seventy-five feet into a shimmering pool. Einarson had seen elk pass through the day before; his partner noted the apparent decrease in deer and speculated that it was due to predation by wolves and cougars, both of which are abundant here. Einarson picked up his train of thought: "Another reason I like falling," he said, "is I like walking around in old-growth forests. It's kind of an oxymoron, I guess—to like something and then go out and kill it." Like a hundred generations of forest dwellers before him, Einarson is also a hunter and a mushroom picker, and in the end he compared his work to hunting: "I've tried taking pictures [of animals], but it's not quite the same because you're not *part* of it."

In this sense, logging isn't so different from the Marine Corps, medical school, or even storytelling: for many of us—even the couchbound readers of books—some sort of blood sacrifice is necessary in order to validate the experience. Of course, any of our lives, closely examined, can be found to hold gross inconsistencies; slaughterhouse workers, loggers, and stockbrokers are simply less insulated from them than the rest of us who benefit from their labors. It seems that in order to succeed—or even function—in this world, a certain tolerance for moral and cognitive dissonance is necessary.

Einarson and his team were cutting a right-of-way for a logging road that would make this remote piece of Vancouver Island accessible to heavy logging equipment. Right behind the fallers was an excavator attended by dump trucks filled with rock for road building, and less than a kilometer behind that was the world's largest known yellow-cedar tree, a massive thing more than a dozen feet across with a trunk covered in shining, velvety moss. Yellow-cedar is the longest-living northwestern tree, and this one could easily predate the fall of Rome. Environmental regulations called for it to be left standing within a tiny set-aside of towering redcedars; Weber and Einarson's boss would later express his regret at not being able to take those trees, too. Nonetheless, in a matter of days, five men and their machines would transform this S-shaped strip of mountain wilderness that included trees nine and ten feet in diameter into a roadway that would be navigable by a grapple yarder, a logging truck, or, for that matter, a Buick sedan. By the time these words are read, the centuries-old cedar, hemlock, and balsam of the cutblock known as Leah Block 2 will be a distant memory, long since processed into siding, two-by-fours, perhaps even the paper that has been recycled into the pages of this book. It will have been accomplished with unprecedented efficiency, but it comes at a price; mechanization is, by far, the leading cause of job loss. The men on this crew can see clearly something their forebears seemed unable, or unwilling, to envision: the end. "It could be argued that we've squandered the resource," observed Weber. "We don't have eight hundred years to replace an old-growth forest. In a few years we'll just have guts and feathers left."

What loggers like Weber and Einarson are seeing on their immediate horizon is a reality that their counterparts in Washington, Oregon, and Northern California are already living with. Collectively, these states have lost more than 90 percent of their old-growth coastal forest, while British Columbia, which originally had twice as much forest area, has lost 60 percent. West Coast loggers, who often find themselves at odds with local Indians, have more in common with eighteenth century Nuu-chah-nulth, Tsimshian, and Haida than they might imagine: while extremely well suited to their environment and the traditional tasks required to survive in it, they are poorly equipped to do much else. Many loggers go into the woods before finishing high school, where "somewhere between a boy and a man," as Weber puts it, "you're making a man's wage." Like the Haida who rode the heady wave of the otter trade, these men found themselves in a situation that is nearly impossible to resist: here you are with a skill set that anywhere else would condemn you to a life of menial labor, and suddenly you're prospering—pulling down fifty or a hundred grand a year in a rural area where living expenses are extremely low. But now these able men with their fantastic machines are racing to the finish line, and praying that they don't get there before it's time to retire.

Just as the Haida were reduced to subsistence hunting, fishing, and potato farming after the crash of the otter population, many West Coast loggers have—in about the same period of time—seen their incomes soar to heights commensurate with a physician's—and plummet to those of a school bus driver, or nothing at all. In this sense, Weber, Einarson, and their long-dead native counterparts are all expendable canaries in the coal mine of resource extraction. When it's finished, these latter-day Nor'westmen will sail away, too, while their wealthy foreign backers search for the next big thing. Out here, the otter trade of tomorrow is oil and natural gas (over the past fifteen years, Prince Rupert has lost approximately 25 percent of its population due to downturns in fishing and forestry). Today, as when the sun touches the horizon, you can almost see the industry's hurtling descent. Even for someone used to the frenetic velocity of urban life, the speed with which this latest logging road unspools across the mountainside and into this rare and lovely corner of the country is sobering. It is like watching the accelerating effect of time-lapse photography on a blooming crocus or a rotting apple, only on a landscape-altering scale.

WHILE THE LOGGING of old-growth forests in Montana, Idaho, Alberta, and British Columbia's interior continues at a rapid rate, the temperate rainforests of the Pacific Northwest represent the end of the line. With the exception of the stunted woodlands of the far north—the boreal forests of Alaska, northern Canada, and Scandinavia, and the taiga of Siberia—there is nowhere new to log in this hemisphere. Anyone looking to China for virgin territory will be sorely disappointed: "Soil and water return to their rightful places," pleads a Chinese prayer attributed to the second millennium B.C., "green colors return to grass and trees." Most people alive today will witness the end of old-growth—big tree—logging, an industry that has been practiced continuously and with undiminished zeal in the Northern Hemisphere for at least five thousand years. By some strange quirk of fate, the largest trees the world has ever known were saved for last—for us.

As paradoxical as it may seem, the fact that West Coast old growth won't be seen again outside of a park for centuries—if ever—is just fine with many commercial loggers. To them, these trees are worth more dead than alive. Out here the frontier age has not yet ended; Weyerhaeuser is not just in the wood products business, after all, they also sell real estate in the form of raw land, freshly cleared for settlement. Within the industry, this practice is known as "log it and flog it," and it is a short-term investor's dream: the landowner gets to liquidate his property not once but twice—first for the wood and again for the land—no need for expensive and time-consuming replanting or stewardship.

Gordon Eason is a senior manager and head engineer at Weyerhaeuser's (formerly MacMillan Bloedel's) North Island Division on Vancouver Island. In addition to being a highly respected forester, he is locally famous for having found the "Carmanah Giant." After hearing an old timber cruiser's story of a huge Sitka spruce growing somewhere in the Carmanah Walbran forest at the south end of Vancouver Island, Eason set out to find it. Since the old cruiser's directions were vague and the Carmanah Valley is vast, Eason flew over the area in a helicopter. Whenever he saw a treetop higher than the rest, he would ask the pilot to hover while he hung out of the door and took a measurement by dropping a logger's tape weighted with a bolt. Most of the taller trees he measured tended to be in the 250-foot range—an impressive height for any West Coast species. As it turned out, these weren't even close. When Eason let the tape go over the Carmanah Giant, the bolt didn't hit the forest floor until it had registered over 300 feet—the same height as the Statue of Liberty, making it Canada's tallest surviving tree, and the world's tallest Sitka spruce.

Eason has spent his entire working life in the logging industry. "I like spending time in the woods," he explained; "that's why I got into it." His only complaint with his current job is that it doesn't allow him enough time to be in the forest. In addition to having the highest density of mountain lions in North America, his territory represents one of the richest remaining reserves of big old-growth timber. By Eason's rough estimate, based on an average annual cut of a million cubic meters, the "allowable" old growth remaining in his region will be gone in thirty-five years. However, it is important to note that all coastal old growth does not necessarily fit the stereotypical image of broad trunks and skyscraping tops. Particularly on mountainsides, old trees tend to be smaller due to shorter growing seasons and poorer soil. The most impressive trees tend to grow at lower elevation and in the best soil—valley bottoms, etc. These areas also happen to be the easiest to log and most of them already have been. Therefore, the 25 million cubic meters that Eason believes will be spared due to inaccessibility or environmental restrictions is likely to be some of the poorest quality—from both an aesthetic and an economic point of view. Furthermore, there is no reason to suppose that Eason's estimated rate of cut won't fluctuate over time, depending on the timber market and changes in harvesting policies and practices. In any case, it is almost certain that technological advances will enable cutting to proceed even more rapidly and efficiently than it already does.

STILL, SOME THINGS never change: in spite of the enormous power and influence wielded by big companies like Weyerhaeuser and Canadian Forest Products (CANFOR), the timber industry continues to ride the same roller coaster of boom and bust that it always has. Wars, prosperous times, and urban catastrophes like earthquakes and fires all signal banner days for the industry, while recessions, depressions, market gluts, and international tariff disputes result in mass layoffs and mill closings. Meanwhile, old-growth forests continue to be viewed with the same combination of awe, appreciation, greed, and contempt that they were in William Bradford's and Plato's day. In the timber industry these ancient woods are known as "decadent forests" because their days of rapid growth are long past and rot is often present—two reasons the industry is in such a hurry to get rid of them. Gordon Eason summed up the prevailing attitude with the same battle cry that loggers—and leaders—have been using for the past five thousand years: "Get that old shit off the landscape so I can get a decent crop out there!"

This is a more complex statement than it appears. It is intended, in context, without any particular malice; it stems, rather, from an unsentimental pragmatism. In fact, it's no different than one of us driving past our local hardware store, as quaint as it may smell and as knowledgeable as its silver-haired proprietor may be, to get to a Wal-Mart. Most of us are led to believe we have more freedom and choice than ever before when in fact we are driven by the real, if short-sighted, demands of our wallets, sophisticated advertisers, increasingly large and powerful conglomerates, and a reactive response to the clock. In this way, tree farms and big-box stores have a lot in common: what they lack in long-term character, beauty, or "soul," they gain in alleged efficiency and cost-effectiveness. It is a side effect of capitalism, the roots of which reach down into our collective attitudes toward nature and the life cycle.

In the modern forest as in the modern retail outlet, the emphasis is—now more than ever—on volume and speed. The "crop" Eason is referring to isn't hay or corn, as it would have been a century ago, but trees—planted in tidy rows, and often in stands of single species rather than the mixed forests that Nature prefers. These are the real biological deserts. Today, trees are bred for speed and are harvested on tight rotations of twelve to eighty years, which is, depending on species and region, the period of the most rapid—read: short-term-investment-friendly—growth. These small, easy-to-manage farm trees tend to be of inferior quality (ask any woodworker); they are often pulpy and loose-grained, and many of them will never be milled into boards at all.

This is the future of the world's "working" forests: a predictable supplier of genetically modified fiber. Increasingly, houses, furniture— the things that form our personal landscapes—are being made not from wood, exactly, but from wood "products": trees that have been ground into chips and sawdust and reconstituted by various means into boards, sheeting, and architectural features. These products have names like Finger-joined lumber (smaller pieces of wood fitted together into boards); MDF (Medium Density Fiberboard); OSB (Oriented Strand Board); WB (Wafer Board); Com-Ply (veneer backed with strand board); cementitious wood fiber (shredded wood that is bonded with cement and formed into boards); Homosote (construction-grade cardboard); Celotex (a variation); Hardboard (a.k.a. Masonite, a denser, thinner version of Homosote); particleboard; and, of course, plywood, which has been around for nearly a century now. The upside is that these products are light, cheap, easy to work with, and make certain kinds of waste a thing of the past, though another kind of waste has been rampant. Millions upon millions of board feet of old-growth Douglas fir have ended up getting peeled for plywood and similarly large quantities of Sitka spruce have been pulped for newspapers and telephone books. This is hard to imagine when one considers that, today, a two-by-ten-inch plank of clear (knot-free) fir or spruce is a luxury item, which perhaps is what it should have been all along. We are slowly being forced to come to grips with the true value of wood, and it's expensive stuff.

If you were to ask a logger today, What is the true value of wood? he would probably answer, "About a hundred and fifty bucks a cubic meter." But a Vancouver building contractor named Duncan Schell penetrated to the heart of the question by responding with an oxymoron. "Wood is priceless," he explained, "but only because it's so cheap." It may sound funny, but this is how our species has collectively evaluated this extraordinary substance that has been central to our survival and success ever since we first picked up a stick. However, Schell's wisdom—so true for so long—is now being put to the test. Fifty years ago, magnificent trees were hauled out of the forest for a pittance, or simply cut and left to rot for the most whimsical of reasons. Today, logging companies are harvesting far lesser specimens with helicopters that cost $5,000 an hour to operate. Fallers can now be seen donning climbing harnesses and rappelling down cliff faces in order to get at previously inaccessible old-growth trees.

According to the Washington Contract Loggers' Association, the average American uses the equivalent of a log 100 feet long by 18 inches across (approximately 235 cubic feet) each year. The amount of wood required to produce an edition of the Sunday *New York Times* would fill the Haida Brave more than one and a half times (almost a million cubic feet of wood).* But even though our appetite for wood is enormous, it has been outstripped by even more rapacious forces. As the planet warms, twin plagues of fire and beetle infestation are laying waste to northwestern forests faster than loggers ever could. By the end of June 2004—early, as far as northwest fire seasons go—a thousand forest fires had been reported in B.C. alone. These, combined with hundreds of other fires burning between Idaho and Alaska, generated a smoke plume that was visible from the Bering Sea to New York City. Meanwhile, the mountain pine beetle has been surviving recent winters in unprecedented numbers and is multiplying at exponential rates. As of 2005, 10,000 square miles of B.C.'s interior forest were infested (an area roughly the size of Pennsylvania), and that number could easily double by the end of 2006. An infested tree is generally dead within a year; if it is not logged within a few years, it loses its value as a source of lumber. After that it can be salvaged only for pulp, if anything. Left standing, it becomes fuel for more fires, which—in the absence of cold winters— are Nature's most effective means of controlling beetle infestations. As a result of these mass forest deaths, "fire sales" of bug-killed and fire-damaged timber are being held throughout the interior Northwest, increasing annual allowable cuts while reducing to "salvage rates" the province's stumpage fees (the per-stump premium paid to the government, which is a key source of income for timber-producing states and provinces). It is a boon to forestry workers, if not to the actual market price of wood, which generally decreases in the face of gluts.

---

*According to the New York Times Company, about 25 percent of their paper is derived from recycled material.

OUT IN HAIDA GWAII, the rain keeps most fires at bay and coastal timber is far less susceptible to the bug infestations that are devastating the interior. It is humans and the things they carry with them that remain the greatest threat to the islands. A terrible irony is that, philosophically, Hadwin was in sync with much of the local population: in December of 2000, an interracial group of islanders staged a protest—essentially, a no-confidence vote—against the Ministry of Forests' handling of logging in the islands. There hadn't been a demonstration of that kind in a decade, and this one was the biggest ever: 20 percent of the islands' adult population participated. Since then there have been some striking changes, not just in the way logging is practiced but in the status of the islands themselves.

Neither the Haida nor any other tribes on the west coast of Canada signed comprehensive treaties with the British or Canadian governments when their lands were first colonized.* A number of tribes are currently negotiating land-claim settlements with the Canadian government, and they are headachingly complex agreements which may ultimately resemble onetime payments of cash, land, and/or percentages of local resource revenues. In 2003, the provincial government made the Haida an offer of 20 percent of the islands and their revenues, but the Haida rejected the proposal out of hand. The tribe has made it clear that it will settle for nothing less than Haida Gwaii in its entirety, including fishing and mineral rights to the surrounding waters. This isn't new; after formally withdrawing from Ottawa's comprehensive land-claims process in 1989, the Haida threatened to issue their own passports. "We have absolutely no intention of ever selling Haida title to Haida Gwaii," said former council president Miles Richardson to a journalist at the time. "We are not, as a nation, going to go cap in hand to any people."

---

*Some tribes did sign a handful of limited, local treaties granting rights to specific coal mines and other resources.

As far as this goes, little has changed in two hundred years. The only difference is that ever since the Haida (along with most other North American tribes) lost control of their historic lands, food sources, and personal destiny, they have been subsidized by the federal government. While subsistence hunting and fishing still play a major role in the lives of the Haida, unemployment—in the European sense of the word—hovers around 80 percent—about the same as in the Gaza Strip. In spite of this, few tribes have the media savvy and charismatic appeal that the Haida do. As grim as some of their demographic statistics are, the Haida are a potent political and social force. This is an amazing accomplishment, particularly when one considers that the Haida are resurrecting themselves much the way botanists have attempted to resurrect the golden spruce. On a regular basis they perform large, inclusive ceremonies whose grandeur, complexity and sheer spiritual voltage is simply stunning. The healing and bonding power of these events is deeply felt—even by off-island visitors.

In 2002, the Haida won a landmark case which required Weyerhaeuser to consult with the tribal council before logging particular areas.*† One result of this is that the annual allowable cut for the islands has been reduced by roughly half, but rather than alienating local Anglo loggers, the Haida have been forming alliances with them. The Anglo residents of Haida Gwaii have spent generations on the front lines of the timber and fishing industries, and they have few illusions about the stated good intentions of powerful entities from off-island. Unlike many loggers, who fly into remote forests and then move on when the trees are gone, most of the residents of these distant, close-knit islands are in it for the long haul; they have nowhere else to go. In 2004, the Anglo residents of New Masset and

---

*The decision was upheld by the Supreme Court in 2004, but the duty to consult was shifted to the province.

†In 2003, Starbucks, the international, multi-billion-dollar coffee conglomerate, was forced to abandon a copyright infringement suit against Haida Bucks Café, a tiny restaurant in Masset, after facing an unanticipated wave of negative publicity and boycotts from around the world.

Port Clements threw in their lot with the Haida, signing an accord that says, essentially, that they trust the stewardship of the local Indians more than that of Weyerhaeuser and the provincial government. Like the logging consultation clause, this is unprecedented in the history of North America. One of the signatories is Dale Lore, the current mayor of Port Clements; a logging road builder by trade, he, like many others, had a revelation in the woods. "I started out as a redneck logger," he told a journalist shortly after signing the protocol affirming the Haida's title to the islands in March 2004. "You know how to beat that picture of a clear-cut in your head? You talk about jobs, that it'll grow back. . . ." But the same questions that tormented Hadwin kept intruding: "What are we getting out of it, what are we doing for the future?" he wondered. "I can beat the picture; I can't beat the epilogue." There is some strong local opposition to Haida title, particularly in Queen Charlotte City, the government hub of the islands. "It's not easy," sympathizes Lore; "the unknown is scary." But then he concludes with what sounds like a page from Hadwin's book: "This is happening because the status quo is obviously fatal to us. People do not change willingly."*

THE FATE OF HAIDA GWAII represents the fate of the Northwest Coast in microcosm, and one of the most extraordinary things about these islands—and much of the North American mainland, for that matter—is how forgiving it is in the face of abuse. Unlike the desertified tracts of the Middle East, this continent—so far—possesses a tremendous capacity for regeneration. In New Eng-land, the cradle of the North American logging industry, remarkable changes

---

* In 2005, Brookfield Asset Management (formerly Brascan), international assets management corporation based in Toronto, took over Weyerhaeuser's coastal logging operations, including those in Haida Gwaii. Many felt that the sale violated the 2004 Supreme Court ruling requiring the province to consult with the Haida before such transfers, and this prompted a month-long blockade of logging operations by Islands Spirit Rising, a coalition of Haida and white islanders, many of whom are loggers.

have occurred as many farmers' fields, which were abandoned after World War II, have reverted back to a forested state for the first time in centuries. In much of the region, the local fauna had long since been reduced to a suburban menagerie of squirrels, chipmunks, groundhogs, and raccoons; thirty years ago, even deer and fox were a novelty. Over the past few decades, however, all that has changed; with the resurgence of the forests coupled with a parallel decrease in hunting, long-banished species have cautiously returned. Coyote, beaver, and wild turkey are commonplace now; the bald eagle is back, too, along with a well-documented explosion in the deer population (which poses a threat to native plant species). If this trend is allowed to continue, it is only a matter of time before the black bear, bobcat, mountain lion, and wolf reclaim their rightful places in New England's long-altered ecosystem. The rivers of the Northeast are another matter: the Atlantic salmon population, in its wild form, has fallen by nearly 75 percent in the past twenty years. Today, the species exists primarily as a farm-raised caricature of itself whose flesh must be dyed pink in order to make it look "real."

Three thousand five hundred miles away, at the far end of the logging continuum, Haida Gwaii's face a much more complex recovery scenario. While the Anglo population has decreased by more than 10 percent in the past decade due to lost fishery and forestry jobs, the native population is resurging. Meanwhile, plans to reintroduce the sea otter are stymied continually by fishermen and abalone hunters who resent the potential competition, despite the fact that it is humans who have devastated the islands' once-abundant abalone. Further complicating matters is a recent proposal to lift a thirty-year-old moratorium on oil exploration around the islands. Ashore there is another major quandary: shortly after the last Dawson's caribou was killed in 1908, Sitka black-tail deer were introduced to the islands; with no natural predators, their population has grown exponentially and they now number in the tens of thousands. No one anticipated that two of their favorite foods would be staples of the understory: red cedar seedlings and salal. Compared to a century ago, many of these islands now have a parklike feel: there is no brush; you can see for hundreds of feet ahead of you. It's beautiful, but the dearth of young cedar is alarming. Cedar has housed, clothed, and defined the Haida for millennia; now carvers are wondering where the next generation of poles is going to be found. The Sitka spruce is doing somewhat better; in the Yakoun Valley, clear-cuts replanted in the 1960s have already grown into forests of 100-foot trees. However, some islands and mountainsides still look as if they have been skinned alive due to the severe erosion that followed the clear-cuts. It remains to be seen whether this new generation of planned forests will ever achieve the elegant and massive complexity of their wild forebears, or if the people who ultimately control them will have the patience and desire to find out.

Revival

How like something dreamed it is.
How long will it stand there now?
　　　　—W. S. Merwin, "Un-chopping a Tree"

Port Clements has suffered much; not only did the town lose its mascot (the golden spruce is the centerpiece for the town logo), but in November of the same year, its albino raven died in a blinding flash when it was electrocuted on a transformer in front of the Golden Spruce Motel. True albino ravens—as opposed to gray or mottled —are all but unheard of. To get an idea of just how rare these birds are, consider this: Alaska and B.C., together cover nearly a million square miles and contain the continent's largest populations of ravens, and yet never in the history of bird observation and collection has a true albino ever been reported in Alaska. Likewise, the Port Clements specimen is the only one ever to have been observed in B.C. (it has since been stuffed and is now on display in the town's logging museum). The raven is the most powerful creature in the Haida pantheon; it was Raven who ushered the first humans into the world. According to one famous Haida story, he started out white, only turning black when he flew out of a bighouse smoke hole, having stolen back the light for a world that had been darkened by a powerful chief. In a strange example of mythical consistency, the white raven's mode of death caused a blackout in Port Clements. Why two unique and luminescent creatures would occur simultaneously against fantastic odds, only to die in such bizarre ways on the same remote island within a few miles and months of each other, is anybody's guess. Science and the mathematics of chance fall short here, so myth, faith, or simple wonder must fill the void.

For most people in the islands, the golden spruce is a fond, sad memory; people who have lost someone dear to them often speak of a light going out in their lives, and so it was with the golden spruce, its loss felt all the more keenly because it had grown in a place where light is such a precious commodity. "It rains a lot here," explained one longtime resident, "and it's cloudy; the golden spruce always looked as if it had the sun on it."

Beneath the scar tissue of forgiveness and philosophical resignation, though, there lies a lingering bitterness that is as pointed as ever. During a meeting with some Tsiij git'anee elders in which they were speculating about the current whereabouts of Hadwin, it became clear that all of them think he is still alive. When one of them suggested that he might come back to the islands, the eldest of them all, a sweet crocheting octogenarian named Dorothy Bell who is known as "the mother of everybody," shook her head. "If he does," she muttered in a baleful tone, "I hope they hang him by his damn neck." This was five years after the tree had been cut down.

During a similar discussion about Hadwin between a group of tugboat operators, one of them, who had unknowingly crossed Hadwin's path in Prince Rupert Harbour, said, "I'd have run him over in my tug if I'd known it was him." Nobody was laughing. The same sentiment was expressed by Dale Lore when a heavy-equipment mechanic named Don Bigg abducted a young Haida woman in December of 2000. After being apprehended and charged in the Masset courthouse, Bigg was put in handcuffs and flown to Prince Rupert in a seaplane along with several other passengers, including the judge who had just heard his case. Halfway across Hecate Strait, however, he decided to exit the aircraft with a 110-pound police escort clinging to his leg. In the end, Bigg went out alone, falling 5,000 feet into heavy seas. His body was never found, but within a week a short, dark joke was circulating: "Hope the bastard landed on Grant."

Relatively speaking, most people up here feel about Hadwin the way people in the States feel about Timothy McVeigh: he's an outsider who came into their place and killed something precious. If they catch him, he will pay. As far as many Haida are concerned, Hadwin is one more white guy who came out to their islands in order to take something away, only to leave behind yet another imported illness: this time, a new strain of terrorism. Hadwin has paid dearly, though; whether he is alive or dead, he has, for all practical purposes, become what the Haida call a gagiid. The word gagiid (ga-GEET) translates, literally, to "one carried away," and it refers to a human being who has been driven mad by the experience of capsizing and nearly drowning during the wintertime. Dance masks depicting this creature are notable for their wild, piercing eyes, and for their blue or green skin, indicating prolonged exposure to cold water. The cheeks are sometimes shown studded with sea urchin spines—a graphic demonstration of the lengths to which the gagiid will go to keep from starving to death as he caroms between worlds in a state of violent and solitary limbo. However, with the right equipment, and proper observance of ritual, the gagiid can be captured and restored to his human state, much as Europeans might treat a traumatized or mentally ill person with love, therapy, or medication.*

---

*The grandfather of the artist Robert Davidson described the gagiid as "a person whose spirit was too strong to die." Says Davidson of the Haida people, "In this way, we are all gagiids."

Ian Lordon, the journalist who covered the golden spruce story for the *Observer*, and whose reporting did the most to reveal its nuance and complexity, understood that history was being made on two levels. "We're witnessing a new Haida story," Lordon explained: "The Death of the Golden Spruce. In a way, we're fortunate to witness an occurrence that was worthy of setting this process into motion."

FOLLOWING THE TREE'S DEATH, a number of ideas were conceived for honoring its memory. Among those suggested were: carving the tree into a totem that would stand vigil over the Yakoun; dividing the trunk into smaller sections which would be distributed among prominent Haida artists for their own interpretations, and milling the wood for guitars. Sitka spruce is one of the best woods in the world for acoustic guitar tops, and a plan was hatched to supply a group of Haida luthiers who were already manufacturing high-end acoustic guitars with wood for a special "Golden Spruce Edition." Some of the reasons none of these ideas got off the ground were the logistics of moving such a big tree out of a roadless fragment of virgin forest, the fact that spruce is much harder to carve than cedar, and human nature in the form of inertia, internecine disagreement, and simple respect for the dead.

In the meantime, the golden spruce has taken on a life of its own—in fact, many lives; it has, in its turn, become a nurse log. Today, the trunk is covered in a thick fur of young seedlings, each one with every intention of beating the odds. But the tree's regenerative powers are also manifesting themselves in some far more surprising ways. In a remarkable feat of adaptation, this tree has harnessed the same species that killed it and made them a vehicle for its own success. Unbeknownst to anyone at MacMillan Bloedel, UBC, or in the Queen Charlottes, the golden spruce has become the most widely dispersed Sitka spruce on earth. And all because of one man.

One afternoon in the spring of 1980, a high school science teacher named Bob Fincham pulled into his driveway in Lehighton,

Pennsylvania, where he found a large box sitting by his garage door. Its Canadian return address was unfamiliar to him, but Fincham, an avid conifer collector, is an optimist, and he opened the package with high hopes. Inside were several plants in plastic gallon pots, and among them was a Sitka spruce. Fincham knows a lot about conifers, and he specializes in cultivariants—aesthetically pleasing mutations bred for garden use—but he had never seen one like this. Nor had he heard of the person who sent it to him: a fellow conifer enthusiast and supermarket butcher from Victoria, B.C. named Gordon Bentham. Bentham, it turned out, was an optimist, too; he had heard of Fincham and his impressive conifer collection, and he was hoping that by sending him one of the golden spruce grafts he had recently acquired, he might get something similarly exotic in return. This unexpected gift was the beginning of a vibrant friendship that lasted until Bentham died in 1991.

Fincham's golden spruce came from the same generation of grafts as those Roy Taylor had acquired (also from Bentham) for the UBC collection, and like them, this one is stunted, plagiotropic, and has never produced cones; other than that, it is perfectly healthy. It even survived a cross-country move to Washington State, where it lives now on the Finchams' new conifer plantation, which includes 1,400 conifer cultivars from around the world. A number of them are golden (there are golden cultivars of many conifer species), but according to Fincham, none of them is as brilliant as the one he calls 'Bentham's Sunlight.' "People see it from a distance," explained his wife, Dianne, "and they want to walk toward it."

In addition to collecting rare and unusual conifers, the Finchams sell them as well, and since Bob Fincham's green thumb extends to grafting, he has been quietly sharing Bentham's Sunlight with the world for more than twenty years. Cuttings of this tree are growing now in Sweden, the Netherlands, Australia, New Zealand, South Korea, and throughout the United States, among other places. The price for a graft in a one-gallon pot is $40, plus shipping. But competition has been heating up recently, and one of Fincham's beneficiaries, Collector's Nursery of Battleground, Washington, has posted the following ad on their Web site:

PICEA SITCHENSIS 'BENTHAM'S SUNLIGHT'–FRESH GRAFT $20.00

NEW! A piece of history from a legendary 300 yr. old Golden Sitka Spruce growing wild on fog shrouded Queen Charlotte Island in Canada, sacred to the Haida Indians, with a tragic end. In 1997 a protester felled this tree in protest to general apathy towards clearcutting. He disappeared before he made it to his court appearance, presumed dead, with only the remains of his broken and battered kayak to be found, and some rudimentary camping gear. A story that has it all—history, sacred symbolism, tragedy, mystery. Grafting material was taken from the downed tree and efforts have been made to graft on to the original rootstock. Read the full story in the American Conifer Society bulletin, fall 1997.

Fincham is a recognized conifer expert and is working on a revision of Krussmann's *Manual of Cultivated Conifers* (Timber Press), one of the standard reference texts on the subject. Unless someone objects before the new edition goes to press, the golden spruce will be included under the name or "epithet" 'Bentham's Sunlight.' In the world of horticulture, the person responsible for naming a new plant or cultivar becomes its "author," and as it turns out, Oscar Sziklai, the author of *Picea sitchensis* 'Aurea' was using an epithet that had already been taken. There is, in Australia, a cultivar of Sitka spruce with a sickly green color—not gold at all—that already goes by this name. But it, too, is invalid because Latinized epithets for cultivars have not been officially recognized by the International Cultivar Registration Authority—the official arbiter of plant taxonomy—since 1958. That year, a new taxonomic policy was instituted that combines Latin with the author's native tongue; for example, Fincham's: *Picea sitchensis* 'Bentham's Sunlight.' How—or if—the

Haida will respond to this remains to be seen, but they have more pressing matters to attend to, the foremost being how to regain control of the islands they have never formally relinquished.

IN THE SPRING OF 2000, Luanne Palmer's golden spruce grafts were declared ready for transplantation, and they were treated with the kind of care usually reserved for masterpieces and controlled substances. Unaware of Fincham's clones, the Haida had made it clear that no cuttings could be taken unless their distribution remained under tribal control. Their primary concern was not that different from MacMillan Bloedel's forty years earlier: they didn't want the tree, or its branches, to be commercialized or turned into souvenirs by aggressive collectors. The Ministry of Forests agreed to these terms and are holding the cuttings in trust for the Haida in a secure location. Because of where they came from on the golden spruce, there is a good chance that these will be dramatically different from Fincham's second-generation clones.

In 2000, the Haida gave the town of Port Clements one of Luanne Palmer's grafts; it was planted next to a church, in the town's new millennium park, where it may be the safest tree in Haida Gwaii. The spindly knee-high sapling is surrounded by an eight-foot chain-link fence topped with barbed wire. In June of 2001, a small group of Haida elders held a private ceremony during which a second cutting was planted beside the stump on the banks of the Yakoun. Both trees are growing in shady areas where they appear reasonably healthy, with golden needles interspersed among the green. Only time will tell whether they will be plagiotropic dwarfs, as every other artificially propagated golden spruce has proven to be, or if they will live up to the lofty message they carry with them from the golden crown of their mother tree.

Board foot (bf) = 12" x 12" x 1"
Cubic foot (cu. ft.) = 12 bf
Cubic meter (cu. m.) = 420 bf, or 35 cu. ft.
Metric ton — 1 cu. m. wood (average)
Cord = approx. 128 cu. ft.

A highway logging truck can carry twenty-five 50' x 2' logs (approx. 40 cu. m.).

Haida Monarch and *Brave* (log barges) can carry approx. 16,000 cubic meters of cedar, 13,000 hemlock (approx. 430 truckloads)

A typical big log (32' x 9') = 11,500 bf or approx. 60 tons

A typical 2,000 sf home uses nearly 16,000 board feet of lumber and 6,000 square feet of structural panels, such as plywood.

ENDNOTES

PROLOGUE: DRIFTWOOD

3   Kayak find information: Scott Walker, personal communication.

CHAPTER 1: A THRESHOLD BETWEEN WORLDS

9   . . . dumped 95 feet (1,140 inches) of snow . . .: "The National Climate
    Extremes Committee's Evaluation of the Reported 1,140-Inch National
    1998–99 Seasonal Snowfall Record at the Mount Baker, Washington,
    Ski Area." Findings presented at the 57th Eastern Snow Conference,
    Syracuse, New York, 2000, www.easternsnow.org/proceedings/2000/
    leffler.pdf.

9   Information on original range and distribution of coastal temperate
    rainforests based on Weigand, et al.

10 "biological desert": Luoma, 42.

11 It has been estimated that a square meter of temperate forest soil . . .:
    Ibid., 94.

11 Andy Moldenke, an entomologist at Oregon State University . . .: Ibid., 97.

13 "west of west": Lillard, The Ghostland People, 33.

13 "One conspicuous feature of the atmospheric effect . . .": Ibid., 305.

14 Very Wet Hypermarine Subzone: Peterson et al., 4.

14 Twenty-three species of whale live in or pass through the region's

waters: Doug Sandilands, researcher, B.C. Cetacean Sightings Network. Personal communication.

15   "She's a black-hearted bitch . . .": Dalzell, *The Queen Charlotte Islands, Book 2: Of Places and Names,* 152.

19   "I didn't even make an ax mark on it . . .": MacMillan Bloedel, 6.

CHAPTER 2: THE PEOPLE

23   "first-class prospectors, and know all about gold mining. . . .": Downie Report 10/10/1859. Lillard 92–95.

24   "very large and capable of carrying one hundred . . .": "The Haidah Indians of Queen Charlotte's Islands," James G. Swan, 1873. Lillard, 121.

25   There has been a great deal of speculation about how far the Haida traveled . . .: See *Tilikum: Luxton's Pacific Crossing—Being the Journal of Norman Kenny Luxton, Mate of the Tilikum, May 20, 1901, Victoria B.C., To October 18, 1901, Suva Fiji.* Sidney, B.C.: Gray's Publishing Ltd., 1971.

25   (Based on existing trade routes and maritime technology . . .): Lillard, *Just East of Sundown,* 51.

26n   While attributed to the Haida, this canoe was probably . . .: Bill McLennan, Curator/Project manager, UBC Museum of Anthropology. Personal communication.

28   "Islands Coming out of (Supernatural) Concealment": Guujaaw, Nathalie Macfarlane. Personal communications.

CHAPTER 3: WILDEST OF THE WILD

37   a legendary place called Fousang: ff. 9, Derek Hayes. *Historical Atlas of the Pacific Northwest: Maps of Exploration and Discovery.* Seattle, Wash., 1999.

37   "the backside of America": Ibid, 7.

38   "as plentiful as blackberries": From Shakespeare's *King Henry the Fourth,* Part 1, Act II, scene iv. Falstaff: ". . . If reasons were as plentiful as blackberries, I would give no man a reason upon compulsion, I." This phrase was well known to literate Britons in the eighteenth century.

39   the equivalent of about $2,400 today: "Inflation Conversion Factors for Years 1665 to estimated 2013," © 2003 Robert C. Sahr; Political Science Department, Oregon State University.

41 Ice-cube–sized hail would cause birds to drop, dead, from the sky: Michael Scott. Personal communication.

41 "dreary," "inhospitable," "wretched," "savage," "barbarous,". . .: Gibson, 147.

41 "shitting through one's teeth": Ibid., 138.

42 smallpox almost certainly preceded the traders: Acheson, *BC Studies*, 50.

43 "These artists of the northwest could dye a horse . . .": Gibson, 159.

45 "it has often been observed when the head of a nail . . .": Ibid, 155.

47 "I could have kill'd 100 more with grape shot . . .": Boit, 49–50.

49 "Before many years . . .": "Missouri River Journals," Aug. 5, 1843, from Alice Ford, *Audubon's Animals*, New York 1954.

50 "Should I recount all the lawless & brutal acts . . .": Gibson, 158.

50 "Spars of every denomination are in constant demand . . .": Gould, 16.

CHAPTER 4: THE TOOTH OF THE HUMAN RACE

55 "What better way to portray the wealth of our country?". Sloane, 73.

61 . . . the term "firestorm" was coined: Pyne, 204.

61 "an enemy to be overcome by any means, fair or foul": Williams, 12.

62 But the "axe age," as one historian calls it: Klenman.

62 Climax, Demon, Endurance, Cock of the Woods, Red Warrior, . . . : Ibid.

64 "The great size of the Timber and the thick growth . . .": Gibson, 71.

64 "When I stood among those big trees . . .": Andrews, *Glory Days of Logging*. Dust-jacket copy.

64 "I raised my eyes to the sky and could see nothing . ": Gould, 15.

65 "British Columbia is a barren, cold mountain country . . .": Ibid., 95.

65 "the numerous and extensive milling establishments . . . ; MacKay, 8.

66 *British Columbia's Supreme Advantage in Climate,* . . . : Gould, 24.

66 "It makes little difference to the people of western Canada . . .": MacKay, 21.

67 "would be selling to the moon . . .": Ibid., 109.

CHAPTER 5: THE BEGINNING OF THE END

69 "You'd gouge into the ground that deep, too. . . .": Wesley Pearson. Personal communication.

73 One county on the Washington coast . . .: Grays Harbor County; Van Syckle, 65.

75 "How'dya like that?": Aubrey Harris. Personal communication.

CHAPTER 6: A BOARDWALK TO MARS

85 "There were some awful bloody animals . . .": Peter Trower. Personal communication.

88 "a wizard on the pegboard": Truls Skogland. Personal communication.

88 "He was very polished . . .": Tom Lundgren. Personal communication.

88 "You never won an argument with Tom Hadwin": Harry Purney. Personal communication.

91 "Even with his hands in his pockets, . . .": Paul Clark. Personal communication.

92 "the unexpected heaped atop the unforeseen": Fawcett, 55.

CHAPTER 7: THE FATAL FLAW

100 . . . the only man-made object besides the Great Wall of China . . .: Brian Fawcett. Personal communication.

100 the derogatory nickname "Brazil of the North": Greenpeace ad, 1993.

100 . . . replanted and renamed a "New Forest": Fawcett, 4.

101 "I was one of the last people to see these areas . . .": Queen Charlotte Islands *Observer.* 1/30/97: 11.

102 they were "a bottom-tier company; . . .": confidential source.

103 "The essence of the spirituality of the desert . . .": *Sayings of the Desert Fathers*, Kalamazoo, Mich., 1987, xxi.

107 "get the fuck out of here!": Confidential source.

112 "The supply of Sitka spruce suitable for aeroplane . . .: Lillard, *Just East of Sundown*, 129.

113 . . . nearly 30 percent of a cutblock's usable wood . . .: Mahood, 52.

117 9 DAYS WITHOUT AN INJURY: Personal visit, 10/01.

119 "If this was mine, I'd cut it all down . . .": Personal communication.

120 "somehow closer and more alive . . .": Parfitt, 107.

121 A senior engineer for M&B who saw the tree . . .: Jim Trebett. Personal communication.

CHAPTER 8: THE FALL

131 Chain saws have been in development since at least 1905, . . .: "Power Saws Come of Age," *The Timberman*, October 1949, 150.

138 "Upset about the Golden Spruce? . . .": Ian Lordon, Queen Charlotte Islands *Observer*. 1/30/97, 11.

139 "This seems to have opened some kind of wound. . . .": *Globe and Mail*, 1/29/97, A6.

139 "Can there be another Gandhi or Martin Luther King?": Robert Sheppard, Victoria *Times Colonist*. 1/30/97, A14.

139 "When society places so much value . . .": Heather Colpitts, Prince Rupert *Daily News*, 2/12/97, 1.

139 "I considered him misguided, . . .": Confidential source.

139 "a great idea. It was M&B's pet tree, . . .": Confidential source.

141 "unofficially, something could happen to him.": Frank Collison. Personal communication.

141 "If I see him," vowed Tranter, "I'll kill him.": Personal communication.

141 "The consensus," claimed Eunice Sandberg, . . .: *Vancouver Sun*, 1/25/97, A1.

142 "nail his balls to the stump.": Personal communication.

142 One Haida leader also suggested that Hadwin . . .: Confidential source.

142 "whether we should cut a part off the person . . .": *Globe and Mail*, 1/27/97, A5.

143 "it makes me sick; it's like losing an old friend.": Queen Charlotte Islands *Observer*. 1/30/97, 10.

143 compared Hadwin's logic to that of the pro-life activist . . .: Prince Rupert *Daily News*, 1/27/97, 4.

143 "They're making it as nasty as they possibly can. . . ": Queen Charlotte Islands *Observer*. 1/30/97, 11.

CHAPTER 9: MYTH

145 "Whoever did this," said a MacMillan Bloedel spokesman . . .: Victoria *Times Colonist*, 1/25/97, 1.

148 "Even now," wrote Pliny the Elder . . .: "Sacred Groves and Sacred Trees of Uttara Kannada," M. D. Subash Chandran. Chapter from *Lifestyle and Ecology*. Edited by Baidyanath Saraswati. New Delhi, India, 1998.

151 "Talking about Skidegate and Masset . . .": Hazel Simeon. Personal communication.

151 "The island population is now shrunk to not over seven hundred. . . .": Bringhurst, 69.

153 "unimprovable" parents: Frank, 35.

155 Lacking anything in the way of the masks . . .: Merrell video.

CHAPTER 10: HECATE STRAIT

159 "He did wrong," she told a journalist at the time.: *Vancouver Sun*, 1/29/97, A1.

161 Veteran North Coast kayakers tell stories . . .: Stewart Marshall. Personal communication.

162 "The worst thing you can do . . .": Ibid.

163 there would be "no uniforms . . .": Heather Colpitts, Prince Rupert *Daily News*, 2/12/97, 1.

169 his wife described him as "indestructible.": Corporal Gary Stroeder. Personal communication.

169 "the Paris of the Pacific": Hayes, 111.

173 "non-ordinary reality": "Shamanic Healing: We Are Not Alone," *Shamanism*, vol. 10, no. 1 (Spring–Summer), 1997.

173 "I never met a shaman who isn't somewhat psychotic . . .": *Maclean's*, 11/4/96, 65.

174 having a "special role" in the world: Confidential source.

174 "very overvalued ideas about the environment . . .": Confidential source.

175 "Instances of such confusion . . .": *ReVision*, 8(2), 21–31, 1986.

175 "Religious or Spiritual Problem.": *DSM-IV*, Washington, D.C., 1994, 685.

175 Spiritual Emergency Resource Center: www.virtualcs.com/se/index.html.

176 "survive on nuts and berries for six weeks . . .": *The Daily Sitka Sentinel*. 6/10/93, 1.

CHAPTER 11: THE SEARCH

179 "everyone is suspect": Confidential source.

180 Dear Lord, give us one more boom. . . .: Local saying.

182 "whip the people's souls" before battle. (and following account): Lillard, "Revenge of the Pebble Town People: A Raid on the Tlingit."

185 "mortician's wax.": Sgt. Ken Burton. Personal communication.

186 "a perpetual tree": www.EXN.ca (Discovery Channel Web site), 1/29/97.

186 "If you live on the edge of the circle . . .": Merrell video.

188 the second tree was a "male": Collison, 39.

188 "a Haida princess guided us to the tree . . .: Victoria *Times Colonist,*
1/29/97, A10.

188 "I wasn't allowed to come to the public . . .": *Vancouver Sun,* 1/28/97, A1.

190 "Nature," it said, "appears reluctant to duplicate a rare, beautiful mistake.": MacMillan Bloedel, 4.

CHAPTER 12: THE SECRET

201 "amorphous, irregular seismic signature . . .": "Strange Beings," Lynn
Lee, *Spruce Roots,* December, 2000.

204 "All you do is attach two wounds together.": Don Carson, www.EXN.ca
[Discovery Channel Web site], 1/29/97.

CHAPTER 13: COYOTE

212 one of Hadwin's previous bosses was reluctant . . .: Confidential source.

213 "Eight hundred years to grow, and twenty-five minutes to put on the
ground. . . .": Personal communication with unidentified source.

CHAPTER 14: OVER THE HORIZON

216 "Civilization has never recognized limits to its needs.": Perlin, 38.

216 "What is not wanted for any present purpose . . .": Williams, 80.

216 "Man," he wrote more than 140 years ago, . . .: Marsh, 228

217 "Nearly the entire territory has been logged over. . . .": F. Roth, "On the
forestry conditions of northern Wisconsin." *Wisconsin Geological and
Natural History Survey Bulletin* no. 1. Madison, Wis., 1898. (See J. T.
Curtis, *The Vegetation of Wisconsin.* Madison, Wis., 1959, 469.)

218 "lady conservationists": Chase, 70.

221 these states have lost more than 90 percent of their old-growth . . .:
Source: Ecotrust.

221 British Columbia, . . . has lost 60 percent.: Source: Sierra Club B.C.

222 "Soil and water return to their rightful places . . .": From *The Book of
Rites,* quoted in "Brief History of Environmental Protection in China,"
Shijiang Peng, Department of Agricultural History, South China

University of Agriculture, Guangzhou, Guangdong, PR China. (*Research in Agricultural History* 1989 (81):131–165. Transl./ interpreted by Dr. W. Tsao, 1/30/2002. Edited by B. Gordon.) www.carleton.ca/~bgordon/Rice/papers/peng89.htm.

229n Starbucks, . . . was forced to abandon a copyright infringement suit . . .: Lane Baldwin. Personal communication.

230 "I started out as a redneck logger. . . .": Prince Rupert *Daily News*, 3/23/04, 1.

231 . . . the Atlantic salmon population, in its wild form, has fallen by nearly 75 percent . . .: "Status of North American Wild Salmon," Atlantic Salmon Federation, May 2004.

EPILOGUE: REVIVAL

234 "It rains a lot here . . ." Personal communication with unidentified source.

234 "the mother of everybody": Collison, 10.

234 "I'd have run him over in my tug . . .": Gunner Anderson. Personal communication.

235 "Hope the bastard landed on Grant.": Personal communication.

235n The grandfather of the artist Robert Davidson . . .: Related during a talk at UBC's Museum of Anthropology, 10/26/04.

BOOKS/ARTICLES

Acheson, Stephen R. "'Ninstints' Village: A Case of Mistaken Identity." *BC Studies*, no. 67 (Autumn 1985): pp. 47–56.

———. "In the Wake of the ya'aats' xaatgaay [Iron People]: A Study of Changing Settlement Strategies among the Kunghit Haida." *British Archaeological Reports*, issue 711. Oxford, England, 1998.

———. "Ships for Taking: Culture Contact and the Maritime Fur Trade on the Northwest Coast of North America." In *The Archaeology of Contact in Settler Societies*, ed. Tim Murray et al. Cambridge, England, 2004.

Andrews, Clarence L. *The Story of Sitka; the Historic Outpost of the Northwest Coast, the Chief Factory of the Russian American Company.* Seattle, Wash., 1922.

Andrews, Ralph W. *Glory Days of Logging: Action in the Big Woods—British Columbia to California.* New York, 1956.

———. *Timber: Toil and Trouble in the Big Woods.* Atglen, Pa., 1968.

Apsey, Mike, et al. "The Perpetual Forest: Using Lessons from the Past to Sustain Canada's Forests in the Future." *Forestry Chronicle*, vol. 76, no. 1 (January–February 2000): pp. 29–53.

Barman, Jean. *The West Beyond the West: A History of British Columbia.* Toronto, Ontario, 1991.

Bechmann, Roland. *Trees and Man: The Forest in the Middle Ages.* New York, 1990.

Blackman, Margaret B. "Window on the Past: The Photographic Ethnohistory of the Northern and Kaigani Haida." *National Museum of Man, Mercury Series* (Canadian Ethnology Service), paper no. 74. Ottawa, Ontario, 1981.

Blackmore, Stephen, and Elizabeth Tootill, eds. *The Facts on File Dictionary of Botany*. Aylesbury, England, 1984

Boit, John (Edmund Hayes, ed.). "Log of the Union: John Boit's Remarkable Voyage to the Northwest Coast and around the World, 1794–1796." *North Pacific Studies*, no. 6. Oregon Historical Society, Portland, 1981.

Bringhurst, Robert. *A Story as Sharp as a Knife: The Classical Haida Mythtellers and Their World*. Vancouver, B.C., 1999

Carey, Neil G. *A Guide to the Queen Charlotte Islands*. Vancouver, B.C., 1975, 1998.

Caufield, Catherine. "The Ancient Forest." *The New Yorker*, May 14, 1990, pp. 46–84.

Chase, Alston. *In a Dark Wood: The Fight over Forests and the Rising Tyranny of Ecology*. New York, 1995.

Collison, Frank, et al. *Yakoun: River of Life*. Massett, B.C., 1990.

Connell, Evan S. *Son of the Morning Star: Custer and the Little Bighorn*. San Francisco, 1984.

Dalzell, Kathleen E. *The Queen Charlotte Islands, 1774–1966*. Terrace, B.C., 1968.
———. *The Queen Charlotte Islands*, Book 2; *Of Places and Names*. Prince Rupert, B.C., 1973.

Davis, Chuck, ed. *The Greater Vancouver Book: An Urban Encyclopaedia*. Surrey, B.C., 1997.

Dietrich, William. *The Final Forest: The Battle for the Last Great Trees of the Pacific Northwest*. New York, 1992.

Ecotrust Canada. *Seeing the Ocean Through the Trees: A Conservation-Based Development Strategy for Clayoquot Sound*. Vancouver, B.C., 1997.

Fawcett, Brian. *Virtual Clearcut: or, The Way Things Are in My Hometown*. Toronto, Ontario 2003.

Fincham, Robert. "Gordon and the Haida." *Coenosium Newsletter,* vol. 1, no. 1 (Winter, 1998).

Frank, Steven. "Schools of Shame"; *Time* (Canadian Edition), July 28, 2003, pp. 30–39.

Gibson, Gordon, with Carol Renison. *Bull of the Woods: The Gordon Gibson Story*. Vancouver, B.C., 1980.

Gibson, James R. *Otter Skins, Boston Ships and China Goods: The Maritime Fur Trade of the Northwest Coast, 1785–1841*. Montreal, 1992.

Gill, Ian. *Haida Gwaii: Journeys Through the Queen Charlotte Islands.* Vancouver, B.C., 1997.

Gould, Ed. *Logging: British Columbia's Logging History.* Vancouver, B.C., 1975.

Grainger, Martin A. *Woodsmen of the West.* London, 1908.

Grove, Richard. "The Origins of Environmentalism." *Nature*, vol. 345 (May 1990): pp. 11–14.

Harrison, James P. *Forests: The Shadow of Civilization.* Chicago, 1992.

Hayes, Derek. *Historical Atlas of the Pacific Northwest: Maps of Exploration and Discovery.* Seattle, 1999.

Hays, Finley. *Lies, Logs and Loggers.* Chehalis, Wash., 1961.

Herndon, Grace. *Cut and Run: Saying Goodbye to the Last Great Forests in the West.* Telluride, Colo., 1991.

Hindle, Brooke, ed. *America's Wooden Age: Aspects of Its Early Technology.* Tarrytown, N.Y., 1975.

Horsfield, Margaret. *Cougar Annie's Garden.* Nanaimo, B.C., 1999.

Klenman, Allan. *Axemakers of North America.* Victoria, B.C., 1990.

Lange, Owen S. *Living with Weather Along the British Columbia Coast: The Veil of Chaos.* Victoria, B.C., 2003.

Lillard, Charles. *The Ghostland People: A Documentary History of the Queen Charlotte Islands, 1859–1906.* Victoria, B.C., 1989.

———. *Just East of Sundown: The Queen Charlotte Islands.* Victoria, B.C., 1995.

———. "Revenge of the Pebble Town People: A Raid on the Tlingit." *BC Studies*, nos. 115 and 116 (Autumn–Winter 1997–98): pp. 83–104.

Lovtsov, Vasilii Fedorovich. *The Lovtsov Atlas of the North Pacific Ocean (1782).* Translated with an introduction and notes by Lydia T. Black; edited by Richard A. Pierce. Kingston, Ontario, 1991.

Lukoff, D., F. Lu, and R. Turner. "From Spiritual Emergency to Spiritual Problem: The Transpersonal Roots of the new *DSM-IV* category." *Journal of Humanistic Psychology*, vol. 38, no. 2 (1998): pp. 21–50.

Luoma, Jon R. *The Hidden Forest: The Biography of an Ecosystem.* New York, 1999.

Macdonald, Bruce. *Vancouver: A Visual History.* Vancouver, B.C., 1992.

MacDonald, George F. *Haida Monumental Art: Villages of the Queen Charlotte Islands.* Vancouver, B.C., 1983.

MacKay, Donald. *Empire of Wood: The MacMillan Bloedel Story.* Vancouver, B.C., 1982.

MacMillan Bloedel, Ltd. "The Backwoods Baronet and the Golden Spruce." Company newsletter. Vancouver, B.C., Nov. 25, 1974.

MacQueen, Ken. "Blood in the Woods: Logging is Dangerous Work." *Maclean's* (January 19, 2004): pp. 31–33.

Mahood, Ian (with Ken Druska). *Three Men and a Forester*. Madeira Park, B.C., 1990.

Manning, Samuel F. *New England Masts and the King's Broad Arrow*. Kennebunk, Maine, 1979.

Marchand, Etienne. *A Voyage Round the World, performed during the years 1790, 1791 and 1792*. London, 1801.

Marsh, George Perkins. *Man and Nature or Physical Geography as Modified by Human Action*, ed. David Lowenthal. Cambridge, Mass. 1965.

May, Elizabeth. *Paradise Won: The Struggle for South Moresby*. Toronto, 1990.

McCulloch, Walter F. *Woods Words: A Comprehensive Dictionary of Loggers Terms*. Oregon Historical Society, 1958.

Nichols, Mark. "The World Is Watching: Is Canada an Environmental Outlaw?" *Maclean's* August 16, 1993, pp. 22–27.

Parfitt, Ben. *Forest Follies: Adventures and Misadventures in the Great Canadian Forest*. Madeira Park, B.C., 1998.

Perlin, John. *A Forest Journey: The Role of Wood in the Development of Civilization*. New York, 1989.

Peterson, E. B., et al. *Ecology and Management of Sitka Spruce, Emphasizing Its Natural Range in British Columbia*. Vancouver, B.C., 1997.

Pike, Robert E. *Tall Trees and Tough Men*. New York, 1967.

Platt, Rutherford. *The Great American Forest*. Englewood Cliffs, N.J., 1965.

Pojar, Jim, and Andy MacKinnon. *Plants of Coastal British Columbia*. Vancouver, B.C., 1994.

Price, Simon and Emily Kearns, eds. *The Oxford Dictionary of Classical Myth and Religion*. Oxford, England, 2003.

Pyne, Stephen J. *Fire in America: A Cultural History of Wildland and Rural Fire*. Princeton, N.J., 1982.

Raban, Jonathan. *Passage to Juneau: A Sea and Its Meanings*. New York, 1999.

Schama, Simon. *Landscape and Memory*. New York, 1995.

Scott, Grant R. "Some Morphological and Physiological Differences between Normal Sitka Spruce and a Yellow Mutant." Undergraduate thesis for Faculty of Forestry; University of British Columbia. Vancouver, B.C., 1969.

Shearar, Cheryl. *Understanding Northwest Coast Art: A Guide to Crests, Beings and Symbols*. Vancouver, B.C., 2000.

Sloane, Eric. *A Reverence for Wood*. New York, 1965.

Snyder, Gary. *The Gary Snyder Reader: Prose, Poetry, and Translations.* Washington, D.C., 1999.

Swanton, John R. (John Enrico, ed.) *Skidegate Haida Myths and Histories.* Skidegate, B.C., 1995.

Thoreau, Henry D. *The Maine Woods.* Boston, 1864.

Trower, Peter. *Bush Poems.* Madeira Park, B.C., 1978.

———. *Chainsaws in the Cathedral.* Victoria, B.C., 1999.

———. *Haunted Hills & Hanging Valleys: Selected Poems 1969–2004.* Madeira Park, B.C., 2004.

Van Syckle, Edwin. *They Tried to Cut It All: Grays Harbor—Turbulent Years of Greed and Greatness.* Aberdeen, Wash., 1980.

Villiers, Alan. *Captain James Cook.* New York, 1967.

Weigand, Jim, et al. "Coastal Temperate Rain Forests: Ecological Characteristics, Status and Distribution Worldwide." Ecotrust/Conservation International, Occasional Paper Series No. 1. Portland, Ore., June, 1992.

Williams, Gerald R. "The Spruce Production Division." *Forestry History Today,* Spring 1999.

Williams, Michael. *Americans and Their Forests: A Historical Geography.* Cambridge, England, 1989.

Wright, Robin K. *Northern Haida Master Carvers.* Vancouver, B.C., 2001.

Wyatt, Gary. *Spirit Faces: Contemporary Native American Masks from the Northwest.* San Francisco, 1995.

VIDEOS

Voices from the Talking Stick. Todd (Tyarm) Merrell. Personal footage; Archie Stocker.

Page 1:
A : Battle between Captain Robert Gray's Columbia and some Kwaikutls. (photograph by George Davidson, Oregon Historical Society, # OrHi 49264)
B: Tenaktak (Kwakiutl) canoes approaching the beach. (photograph by Edward Sherrif Curtis, reprinted with permission. British Columbia Archives D-08429)

Page 2:
A: Outfit of a north coast warrior. (photograph © Canadian Museum of Civilization, catalogue no. VII-X-1073)
B: Steel war dagger. (photograph © Canadian Museum of Civilization, catalogue no. VII-B-944)

Page 3: Dance mask. (photograph © Canadian Museum of Civilization, catalogue no. VII-B-109)

Page 4/5: Skid Road, ox team. (photograph used with permission of The Vancouver Public Library, VPL 3598)

Page 6/7: "Railroad show." (photograph by Darius Kinsey, Courtesy Critchfield Logging Company)

Page 8: Haida fallers. (13017 Merill and Ring at Pysht Darius Kinsey Collection, photograph used with permission of the Whatcom Museum of History & Art, Belligham, WA)

Page 9: High rigger topping. (Leonard Frank Collection, Vancouver, B.C.)

Page 10: Faller watching for widowmakers. (photograph courtesy of Al Harvey)

Page 11:
A: Feller buncher. (photograph courtesy of Al Harvey)
B: Clearcut. (photograph courtesy of Al Harvey)

Page 12: Grant Hadwin. (photograph by Rudy Kelly)

Page 13: Grant Hadwin embarking by kayak. (photo courtesy of *The Prince Rupert Daily News*)

Page 14:
A: Hyder, Alaska. (author photograph)
B: Leo Gagnon and his son. (author photograph)

Page 15: Mortuary poles. (photograph courtesy of Al Harvey)

Page 16: Memorial pole for Ernie Collison (Skilay). (author photograph)